DIMENSIONS OF APEIRON

A Topological Phenomenology of Space, Time, and Individuation

VIBS

Volume 154

a volume in
Philosophy and Psychology
PAP
Jon Mills, Editor

DIMENSIONS OF APEIRON

A Topological Phenomenology of Space, Time, and Individuation

Steven M. Rosen

Amsterdam - New York, NY 2004

Cover Design: Paul Pollmann

The paper on which this book is printed meets the requirements of "ISO 9706:1994, Information and documentation - Paper for documents - Requirements for permanence".

ISBN: 90-420-1199-8

Printed in the Netherlands

To Marlene and Myra, for their loving support from beginning to end

CONTENTS

LIST OF FIGURES

FOREWORD

Within the history of Western philosophy, the concepts of space and time have traditionally been privileged ontological categories that inform how we conceive of nature and mind, reason and science, self and society, and the lived quality of human experience. In this highly intriguing and perspicacious work, Steven M. Rosen explores the Greek notion of the *apeiron*—the boundless and unlimited, spaceless and timeless aspect of nature that must now be taken into account—within our postmodern world marked by radical plurality, destabilization, and disequilibrium that threaten individual well-being, cultural unity, and cosmic order. For the pre-Socratics, *apeiron* signified the principle of disharmony punctuated by chaotic multiplicity, thereby eclipsing any notion of the autonomous individual subject, a position reclaimed by postmodern currents today. Tracing the emergence of individuality from ancient mythic culture to modernity, the scientific revolution, and technological contemporary society, Rosen cogently shows how the crisis of individuation cannot be resolved without addressing the question and transmutation of space and time in light of *apeiron*.

Rosen provides a robust and maverick rethinking of the space-time conundrum that draws on the developmental progression from ancient philosophy to alchemy, existentialism, physics, topology, and ontological phenomenology. The author succeeds admirably in recapitulating the call to Being as a phenomenological enterprise, namely, the realization of individuality through a concretization of abstraction that pays heed to *apeiron*.

Dimensions of Apeiron is an original, highly thought-provoking, superb treatise on the lived experience of temporal embodiment within the disparate pluralities that define our collective social life. This book is destined to reshape our understanding of this elusive and paradoxical topic for generations to come.

Jon Mills
Editor
Philosophy and Psychology

PREFACE

Change itself changes. Over the last 150 years, it has assumed a qualitatively new, more emphatic form. The continuous orders of change that were characteristic of earlier times have given way to a thoroughgoing *dis*continuity, and this has affected virtually every sector of modern life: popular culture, art, literature, science, mathematics, the media, etc. (see Everdell, 1997). The gradual evolution of our social, political, and economic institutions has been disrupted by the onset of multiple crises that have thrown many of these systems into disarray. There has been a disintegration of family and church. The ethnic conflicts that rage around the world today reflect a destabilization of national identity. We have seen an alarming growth in international banditry and terrorism. World markets have reached new levels of erratic fluctuation. Nuclear weapons and waste are proliferating out of control....Thus it has gone. As a consequence of all this, "fragmentation is now very widespread, not only throughout society, but also in each individual" (Bohm, 1980, p. 1). Alberto Melucci observed accordingly that, given "the surging flux of events and relations…[t]he points of reference used by individuals and groups in the past to plot their life courses are disappearing" (1996, p. 2).

Taken collectively, these cultural discontinuities constitute the crisis of postmodernity. Just what happened to precipitate the dilemma? Was it a particular event? Was it a set of events occurring in certain places at certain times that led us to this pass? I suggest that what happened was a transformation of our experience of space and time themselves. At bottom, it is the sense of *space and time* that began to be fragmented in the middle of the nineteenth century. The introduction of non-Euclidean geometry, the advent of Impressionist art, the Einsteinian revolution in physics (to give a few preliminary examples) all bear witness to the impending breakup of classical space. Along with this, "[o]ur experience of time undergoes increasing fragmentation....Linear time yields to an experience of transitions without development, to a movement between disconnected points, a sequence of fleeting moments" (Melucci, 1996, p. 9). The old sense of smoothly passing from past to present to future becomes a spasmodic leaping ahead into a boundless array of divergent possibilities. In short, the contemporary experience of space and time entails the "shock of discontinuity"—to cite John Berger's and Jean Mohr's (1982, p. 86) characterization of the shocking effects of photography that were especially evident when the medium was first conceived in 1839.

Assuming that the mid-nineteenth century unsettlement of classical space and time has in fact contributed to our current state of turmoil in a significant way, how can we better understand it? What I will show in this book is that the nineteenth century watershed betokened the reappearance of an old nemesis. Western culture was forged from the struggle of human reason with the irrational forces of nature. To early Greek science and philosophy, nature in the wild is *apeiron*. This is the Greek word for what is "limitless," "boundless" or "indeterminate." The *apeiron* is variously interpreted as "the unintelligible; the many; the moving; the ugly; the bad…the inchoate flux of opposites or contraries…the principle of disorder or disharmony" (Angeles, 1981, pp. 14–15). In its sheer boundlessness, *apeiron* defies containment within the ordering contexts of space and time. To the early Greeks, this posed a considerable challenge. For, in the unconstrained many of *apeiron*, there can be no *one*; in its chaotic multiplicity, there can be no unity, no stable center of identity, no indivisible core of being, no *individual*. So it seemed to the early Greeks. It was therefore imperative for them to tame *apeiron*, given the primary impulse that motivated their action. To paraphrase Protagoras, "man must be the measure of all things." What this basically required was the ascendancy of the autonomous individual. More generally stated, from the outset Western culture has been spurred by the drive toward differentiated being or individuality, toward *individuation*. Achieving this end essentially has meant containing what at first appeared uncontainable: the boundless *apeiron*.

The proposition I submit is that *apeiron*, after being held at bay for over two thousand years, has now returned with a vengeance. This, I suggest, is the dilemma that underlies postmodernity. The disruption of space and time, and, along with it, of all the "points of reference used by individuals and groups in the past to plot their life courses" (Melucci, 1996, p. 2), thus seems to point to a total unraveling of centuries of progress in human affairs. All that we have gained seems threatened; all seems on the verge of being lost. But appearances can be deceiving. What I intend to demonstrate in the chapters to follow is that the upsurgence of *apeiron*—far from indelibly spelling the demise of human individuality, actually offers us the opportunity to bring it to fruition.

1. Preview of the Chapters

The process of coming to terms with *apeiron* in the interest of individuation cannot be separated from the transmutation of space and time. That is a central thesis of this book. In the introductory chapter, I sketch the emergence of individuality from ancient mythic culture to modernity and show how it is integrally coupled with an increasing crystallization of the experience of space as a well-sealed container, a seamless continuum. Then, in the remainder of Part One, we explore in depth the mid-nineteenth century upheavals in science (particularly, in physics), and in the broader culture (the media, art-world, etc.), that attended the resurfacing of *apeiron*. Because *apeiron* was not recognized for what it is, its profound challenge to the old space-time framework could not be addressed effectively. And the denial of *apeiron* is what eventually leads to the crisis we now confront.

Part Two is chiefly concerned with the shift to a fundamentally new understanding of space, time, and individuality, one that entails not only the recognition of *apeiron*, but its radical rethinking as well. Chapter Three paves the way for this transition with a study of its philosophical precursors, as expressed in the existentialist writings of Søren Kierkegaard and Friedrich Nietzsche, the modernist phenomenology of Edmund Husserl, and the post-modernism of Jacques Derrida. With the stage thus set, the ontological phenomenologies of Martin Heidegger and Maurice Merleau-Ponty are introduced in Chapter Four. We now pass from the space-time of modernist science and philosophy to phenomenological *time-space*, a term coined by Heidegger to signify the dimension of ontological process, the embodied action sphere of Being. So the new individuality that accompanies the new sense of time and space does not primarily refer to individual beings with novel characteristics, but to the individuality of Being per se. Later in the same chapter, dimensional Being is likened to a sentient living organism whose stages of growth run a (non-Hegelian) dialectical course involving self-alienation and transformative self-return. In this developmental context, classical science is presently related to Being's self-concealment, and the mid-nineteenth century disruption of science (and of the broader culture) is re-interpreted as a call back to Being.

But was it not *apeiron* that was concealed until mid-nineteenth century, and *apeiron* that then resurfaced? What Heidegger explicitly discloses in his meditation on the oldest fragment of Western thought (the Anaximander fragment) is that Being is none other than *apeiron*. From this it is evident that Being is no principle of "pure reason" like the Logos or God of patriarchal tradition, but possesses a distinctly disorderly side, an aspect that is "brute or wild" (Merleau-Ponty, 1968, p. 170). Chapter Four thus makes it clear that the return to Being that would advance the process of individuation hardly would entail a simple vanquishing of *apeiron's* chaos in favor of perfected order and unadulterated rationality. Yet neither would the return be a mere *regression* to chaos. In Chapter Four we come to see that the initial banishment of *apeiron* by the ancient Greeks was at bottom not the triumph of pure reason over an alien force, but signified instead *apeiron's* concealment of *itself* in the interest of its own maturation. And the reemergence of *apeiron* at mid-nineteenth century is the call back to Being heralding the climactic stage in the work of individuation.

By way of heeding the "call to Being" (Heidegger, 1964/1977), Chapter Five provides a further specification of the ontological dimension, a more detailed, concrete realization of its mode of action. For this I turn to visual geometry and qualitative mathematics, and, in particular, to topology. Since my aim is surely not to *objectify* Being, conventional topology is thoroughly recast along phenomenological lines. But a further step is required in making apeironic Being a concrete reality. It is evidently not enough to write *about* Being, assuming the customary posture of authorial detachment and anonymity. If Being's actual presence is to be secured in the ontological text, rather than merely predicating Being—signifying it in such a way that it is implicitly projected as exterior to the author's semiotic act—the author must signify Being by signifying *himself*. The self-signification of the text is taken up in the final section of Chapter Five.

2. Science and Phenomenology

In addressing the contemporary cultural crisis, I give special emphasis to science, and, in particular, to the discipline of physics. Why is this done?

The existential phenomenologist Karl Jaspers comments that, in the century following the death of Immanuel Kant,

> Science has gained an ever-growing overwhelming importance; by its consequences it has become the fate of the world. Technically, it provides the basis for all human existence and compels the unpredictable transmutation of all conditions....Science cannot be avoided. It extends further than in Kant's time; it is more radical than ever, both in the precision of its methods and in its consequences. (1941/1975, p. 166)

Heidegger comes to a similar conclusion. In introducing his essay on metaphysics (1929/1975), he notes that, "metaphysical questioning…has always to be based on the essential situation of existence" (p. 242). What is the "essential situation" at this juncture in our history? It is that we are "ruled by science" (p. 242). When Heidegger adds that "science has become our passion" (p. 242), the collective personal pronoun does not merely denote certain individuals or groups, or even the bulk of humanity. For, in constituting the "essential situation of existence," the preeminence of science extends to the core of the contemporary human being as such (historical *Dasein*). Therefore, whether or not we consider ourselves to be "scientists," we cannot simply turn away from science since the "we" that would turn is defined essentially by its scientific attitude. Science then must indeed play a central role in any attempt to address the crisis of our times.

But does science itself not *contribute* to the crisis in a primary way? In his essay, Heidegger asks, "What essential things are happening to us in the foundations of our existence, now that science has become our passion?" (1929/1975, p. 242). The answer is clear: We are experiencing fragmentation. Is this not reflected in science's *modus operandi*? Its fundamental method is *analysis*, a word of Greek origin that means a "dissolving, a resolution of whole into parts; *ana*, up, back, and *lysis*, a loosing, from *lyein*, to loose" (*Webster's*, 1976, p. 64). Or we may say equivalently that "analysis" connotes "a breaking up" (*American College Dictionary*, 1968, p. 45). Science thus proceeds by breaking things up, or—to use the more common manner of speech—it breaks things down. In fact, the process actually entails a *threefold* breakdown or division, since the object to be dissected scientifically must first be extracted from its context, and since the analyst him- or herself must assume a detached stance.

However, science's penchant for division is not the whole truth. For, in the process of dividing the objects of analysis into parts, it is essential that the subject, the scientific analyst himself, remains inwardly *un*divided, retains his coherence and integrity. We may say that scientific analysis—whatever the purpose of the particular research—is carried out for the ultimate purpose of enhancing the "indivisibility of the subject," that is, promoting individuality, advancing the goal of individuation. In the humanistic program of classical science, the more effectively the objects can be parted, the better they can be controlled, manipulated, and shaped so as to solidify the *wholeness* (unity, oneness, identity) of the subject. Until the middle of the nineteenth century, this was clearly a viable and fruitful project. The breakdown of objects had taken place within a framework that itself had held together. All objective analysis had been carried out within the integral context of space and time, and this permitted the analyst to implicitly preserve and bolster his own sense of integrality. Then came *apeiron* to radically challenge the classical attitude via its disruption of space and time. We are going to see that the crisis of fragmentation now besetting us certainly did not arise from openly embracing resurgent *apeiron's* chaotic multiplicity, but precisely from science's unsuccessful effort to deny it.

It is true that we cannot simply turn away from science, given that it has established itself at the core of our being. But this is not to say that we must accept science in its presently dominant form. I propose that, while every effort to go *outside* of science indeed will leave us squarely in its midst, science can be profoundly transformed. To achieve this we must move in the other direction. Questioningly, we must go all the way into science—go so deeply into it that we come out on the "other side," so to speak. Of course, this requires that we refrain from merely rejecting science. In the course of this book, the positive side of science will not be neglected. Eventually, we shall see that science is the gift that Being gives to itself by way of fostering its own development.

Elsewhere, I spoke of the need for a *self-deconstruction* of science (Rosen, 1994, p. xviii). This amounts to an "archeological expedition" into the foundations of science in which science is deeply questioned from within itself in the interest of radically transforming itself. Needless to say, there is little appetite for such a venture in conventional science, or in the analytical approach to philosophy that unswervingly adheres to this convention. As Leonard Lawlor puts it in his phenomenological critique, "analytic philosophy still sees itself as a 'chapter' in modern natural science" (1999, p. 27). However, it seems to me that contemporary phenomenology, for its part, harbors in certain quarters a rejectionist attitude toward science that—far from diminishing its influence, in fact only serves to maintain the status quo. To be sure, there are phenomenologists and post-phenomenologists who do not merely seek to avoid science. For example, Husserl (1936/1970) and Heidegger (1927/1962, 1954/1977, 1962/1977, 1964/1977) have written extensively on science, and Michel Foucault (1972) has offered his own "excavation" of it. Why is it that science itself apparently has remained impervious to these critiques?

At one level, it can rightly be said that the critiques have failed to have an impact on science because they have originated outside of the domain they have challenged. If science is to be transformed, detached commentary will not suffice, however perceptive it may be. Science must be engaged concretely. What this requires is close contact with science's own language system and frame of reference—with the specific problems and perspectives of science, as science itself understands them. Those who seek to influence the foundations of science without engaging with the actual work of science will meet only with frustration.

But have phenomenological and post-phenomenological critics of science really gone *outside* of science? How could this be true if the rule of science is all-pervasive, as claimed above? I suggest that, in the deepest sense, the critics actually have not managed to transcend science. Although they indeed have positioned themselves outside of a certain cluster of disciplines and the particular practices in which they engage, by no means have the critics surmounted what is *essential* to science: that posture of detached analysis whereby objects are divided by a subject who would maintain his own indivisibility. Thus, it is not just that phenomenological and post-phenomenological critics have failed to influence science. For, in the very attitude of detached analysis they have adopted toward science itself, they have positively reinforced it.

The basic point is that contemporary science is so influential that it simply has no outer limit, no exterior boundary that could serve as an exit.

But I am proposing that it does have an *inner* boundary. My suggestion is that science can genuinely be transformed—both in its particular disciplinary practices and in its essence—not by attempting to critique it from "afar," but by moving completely inside of it. A radical deconstruction of science can surely occur, but only in the form of a *self*-deconstruction, one that engages science concretely, wholly enters into it. If it is true that one must enter the heart of science to fully transform it, then critics who situate themselves outside of science's core disciplines in fact just guarantee the perpetuation of science in the essential form it presently assumes.

The phenomenological transformation of space, time, and individuality initiated in this book undertakes a self-deconstruction of science, with some of science's most basic problems being taken up in science's own terms. For this I focus on what has come to be known as the "king of the sciences," the discipline of *physics*. It is physics that all other sciences (both natural and social) have adopted as their paradigm. And it is physics that deals most deeply and directly with the structure of space and time. Therefore, if the fragmentation of space and time does underlie today's sociocultural crisis as I have suggested, physics is the ideal medium in which the dilemma can be studied and addressed. Accordingly, Chapter One features a detailed examination of developments in classical and modern physics—including relativity theory, quantum mechanics, and unified field theory. In subsequent chapters, contact with physics is maintained and its relationship to ontological phenomenology explored. However, I must acknowledge that the transformation of physics as such is not actually *completed* in the present volume. Originally, I had envisioned a chapter that would serve this purpose by explicitly directing itself to physics' most fundamental problem: that of unifying the laws of nature. The chapter was to follow up on and carry forward my earlier efforts along these lines (Rosen, 1988, 1994). What I realized in the end is that a single chapter would never suffice; that nothing less than an entire volume would be required for an adequate phenomenological rendering of the unification of physics. To the extent that the current volume does deal with the structure of space, time, and individuality in a concrete and comprehensive fashion; inasmuch as it employs topology, a tool in the repertoire of mathematics, known as the "*queen* of the sciences"—this phenomenological work may indeed be considered a self-deconstruction of science. But to complete the self-deconstruction of the "*king* of the sciences" per se, a whole new volume will be needed. (See concluding paragraphs of Chapter One. For related approaches to the deconstruction of science, see Gendlin and Lemke, 1983; Havel, 1998; Matsuno, 1999; Restivo, 1988; and Schumacher, 1989.)

3. Abstraction, Concretion, and the Self-Deconstruction of the Text

The reader who continues beyond this Preface will embark on a journey that ranges from the heights of philosophical abstraction to the depths of concreteness. The word "abstract" is from the Latin *abstractus*, "dragged away, pp. of *abstrahere*, to draw from or separate" (*Webster's*, 1976, p. 8). There is a clear-cut functional relationship between abstraction and the approach we discussed above: *analysis*, which involves "a breaking up." In both cases, the basic procedure is that of division, with division becoming fragmentation in the end. In fact, the analytic work of "pure science" can be taken as the prime instance of abstract functioning, since mathematics, science's preferred language, carries abstraction to its highest degree. To be "ruled by science" then, is to be ruled by *abstraction*. And what the *self-deconstruction* of science must do is question that rule from within.

Where does an effective questioning of abstraction begin? In its most concrete manifestation, I suggest. In writing about the abstractions of science and philosophy, the most immediate fact is that *this text itself* is abstract. The rule of abstraction clearly is in evidence right here, not just in the content of the discourse, but also, in the very manner in which we are communicating (in my way of writing and your way of reading). This means that, for self-deconstruction to be enacted in this text, the text must deconstruct *itself.* Not only must the abstractions of science and philosophy per se be questioned but also, our way of communicating about them, since that too is ruled by abstraction. The dialectic that governs this process has already been foreshadowed: To get beyond the abstract mode of functioning that we are involved in to a more concrete one, to reknit that which was torn asunder, we must move *into and through* our abstractness. So the radical rethinking of space, time, and individuality that is required in order to confront the problem of postmodern fragmentation will entail an examination of the very manner of communication in which we are presently engaged. By going all the way into this abstract mode of discourse, we eventually will reach its "interior boundary" and "come out on the other side." That is the hope.

4. The Dialectic and Alchemy

In both form and content, the overall organization of this book follows the dialectical path from abstraction to concreteness. The odyssey begins in Part One with an ascent into the rarefied climes of theoretical science. Then, in entering Part Two, we turn the corner and start our "descent." Concrete phenomenological awareness is brought to bear, followed by a visual and topological concretization of phenomenology itself. The process reaches its climax in the concluding chapter, where the gap is closed between the topological embodiment of three-dimensional Being and the body of this very text.

Now, a leitmotif of my presentation is the "pre-scientific" discipline of *alchemy*. Metaphorically speaking, it could be said that the structure of this book coincides with the two main stages of the alchemical opus. The motto of alchemy was *solve et coagula*, "dissolve and coagulate." The term "dissolve" is functionally related to the words "analyze" and "abstract." To "dissolve" is "to disunite; to break up...to cause to separate into parts" (from the Latin "*dissolvere*, to loosen; *dis-*, apart, and *solvere*, to loosen"; *Webster's*, 1976, p. 532). The alchemical substance to be transformed was thus initially broken into parts, sublimated so as to refine it. Then, in the second stage, the process was reversed and the material was made to congeal, its constituent elements being drawn together and solidified. This was the phase of "coagulation" (from the Latin "*coagulatus*, pp. of *coagulare*, to curdle"; *Webster's*, 1976, p. 345). What had been divided was now reconstituted.

Importantly, the alchemical "substance" was not just a material object. The transformation in question was to include the *mind*, human consciousness itself. The stage of "dissolution" basically entailed a dissolving of the mind's ties to the body. The process was one of *abstraction*. And the subsequent stage of "coagulation" involved a *concrescence* in which mind and body were reunited (the word "concrete" is from the Latin *concretus*, pp. of *concrescere*, "to grow together, to be solid"; *Webster's*, 1976, p. 378). As the alchemist Gerard Dorn put it: one must "free the mind from the influence of the 'bodily appetites and the heart's affections'....In order to bring about their subsequent reunion, the mind (*mens*) must be separated from the body...for only separated things can unite" (Dorn cited in Jung, 1970, p. 471).

The headings of the two main parts of this book invoke the alchemical metaphor. Part One is *Solve* and Part Two, *Coagula*. Dare I suggest that these designations may be regarded as *more than* metaphorical? For Part One does focus on the abstractions involved in the rise of modern science, and it is an historical fact that the founding fathers of this movement (Kepler and Newton, for example) were serious students of alchemy. Even though the current conventional wisdom is that modern science wound up *repudiating* its alchemical origins, could this apparent repudiation actually have been an integral part of alchemy itself—a move toward greater abstraction that was necessary for the furtherance of the *solutio*, the stage of dissolution (see Schwartz-Salant, 1998, p. 14)? If there is any truth to this reading, we might then be able to say that the contemporary turn to the concrete work of phenomenology constitutes an entry into the *second* stage of alchemy, the *coagulatio*.

Let me take note of an additional parallel to alchemy. Far from being straightforward, the alchemical process was riddled with paradox (Jung, 1967, 1968, 1970; Rosen, 1995). The relationship between the two main stages of the opus is certainly no exception. It would be an oversimplification to say that, whereas the *solutio* brings the practitioner from concreteness to abstraction, the *coagulatio* returns her or him to concreteness. The transition to stage two is not just a movement *away from* mind and back into the body; rather, it entails a *union* of body and mind (as noted by Dorn), one with a paradoxical twist: mind and body indeed become one, yet also retain their distinctness. In the language of the "Book of Lambspring" cited by the alchemist Johannes Fabricius, "The two...are two, and nevertheless they are one" (1976, p. 190). This peculiar dialectic of identity and difference is equally applicable to the relationship between the two parts of the present book.

The turn to concreteness that occurs in Part Two is by no means just a turning away from the abstractions of Part One. Despite the new emphasis on concrescence, abstraction continues. In fact, we go further into it, take it to its limit. What happens when we reach the boundary? *Do* we finally "come out on the other side," the "side" of full concrescence? Yes, we do. And also—because the boundary in question is an *interior* one—we do not!

Let me attempt to clarify what I mean by an "interior boundary." In conventional thinking, the idea of the boundary plays an indispensable role. That is because the convention essentially entails *abstract analysis*, and this requires the drawing of categorical boundaries to set things apart from each other in a clear-cut way. Under the rule of abstraction, we strictly adhere to the logic of either/or: Either we are abstract or concrete, inside or outside, different or the same, mental or physical, etc. What of the *interior* boundary?

By this term I mean the boundary or limit of *limitative thinking itself.* Abstraction's inner boundary is its natural point of termination, its true end. Yet the true end of abstraction cannot *merely* be an end, since any "clean break" of this sort would only testify to the fact that abstraction was actually still taking place! Like the proverbial Chinese finger puzzle, all efforts to break free of abstraction leave us squarely within it. Why? Because that is what abstraction is all about: the effort to *break free*, to produce clean breaks (to break up, break down, analyze, dissolve, etc.). In order for abstraction truly to end, there is no avoiding paradox—an end that is *not* an end, a boundary that is not one. Thus, while we do engage the concrete in crossing the inner horizon of abstraction, this movement beyond abstraction is at once also a movement within it. Such is the peculiar logic of self-deconstruction. Only by remaining *within* the abstraction of this text can we radically surpass it. Like the movement from one side of a Moebius strip to the other that paradoxically keeps us on the *same* side, our passage from abstraction to concrescence at once maintains the former. (The Moebius strip in fact will play a pivotal role in the topological work of Part Two.) Of course, the *supremacy* of the abstract is *not* maintained. What we realize instead is an internal harmony of abstraction and concrescence in which the prior meaning of each term changes profoundly.

Speaking of "boundaries" and "limitations," it is time for me to acknowledge a notable limitation in the book you are about to read. To realize my agenda within the covers of a single volume, it has been necessary for me to paint with a broad stroke. The chapters that follow span a multitude of disciplines: philosophy, physics, mathematics, history, religion, psychology, anthropology, art, and others. Given the scope of the work, in no way would it have been possible for me to operate at the finely calibrated level of scholarly detail that the specialist can achieve in more tightly circumscribed areas of investigation. Still, I have made every effort to remain faithful to the facts that have been established in the fields I have broached. Beyond the criterion of consistency with external fact, it seems that a work of the present kind above all needs to be judged in terms of its *internal* consistency and its general coherence. On this count, validation is contextual; an idea gains credibility and weight to the extent that the integral role it plays within an overall network of interrelationships is plain to see. Since the clarity of the whole is at least as important as that of the separate parts in the present book, since the parts are illuminated by the whole, my hope is that I have been skillful enough and passionate enough to make the whole transparent.

However, there is no denying that the agenda I have laid down is difficult, to say the least. In rethinking space, time, and individuality from the ground up, we will need to think dialectically, engage in hybrid blendings of thought that transgress long-cherished categories, indeed, that fly in the face of categorial thinking as such. Why should we bother? Because it is the only way to deal with the fragmentation presently tearing us apart. That is my conviction. While the path I have outlined is fraught with daunting challenges, I believe and will attempt to show that there is no other effective way to come to terms with the crisis of postmodernity. The only path is through the "eye of the needle." The only way out is all the way *in*.

In closing this Preface, I would like to acknowledge the support I have received from a number of individuals in the course of preparing this book. For their thoughtful comments on preliminary drafts of the manuscript, I am gratefully indebted to David Dichelle, Joel Funk, Eugene T. Gendlin, Lloyd Gilden, Walter Glickman, Marketa Goetz-Stankiewicz, Brian D. Josephson, David Lukens, Koichiro Matsuno, Lael McCall, Yair Neuman, Rod Pennett, Milan Pomichalek, Marlene A. Schiwy, Ernest Sherman, W. J. Stankiewicz, Louise Sundararajan, Geo N. Turner, and John R. Wikse. Much appreciated is the encouragement and helpful feedback that came from Jon Mills, the

Associate Editor who first welcomed me to the Value Inquiry Book Series. For their patience and generous assistance with the technical preparation of this monograph, I give my thanks to Eric van Broekhuizen, Managing Editor for VIBS, and to Shelley MacDonald, Rachel Moffat, and Mark Lewental. And I want to thank Martin Gardner and Paul Ryan for their kind permission to use their topological drawings in Chapter Five of this book.

5. Postscript on Gender

The gendered nature of the quest for individuation is undeniable, and there is a vast literature on the subject (Cixous, 1980; Wolf, 1984; Irigaray, 1985; Eisler, 1987; etc.). In the present volume, the gender factor will become evident in due course. For the moment, let me just note that my use of personal pronouns is informed by this consideration. In dealing with activities that reflect the dominant influence of patriarchy, such as the work of classical science, I normally use the masculine personal pronoun. The feminine personal pronoun is employed when we come to recognize the gender implications of Being ("she" is "motherly" or "grandmotherly"—more like a goddess than a god). In all other cases, I use an assortment of gender-neutral expressions: "his or her," "s/he," alternations of "he" and "she," and so on.

Part One

SOLVE
THE FLIGHT FROM *APEIRON*

One

THE RISE AND FALL OF CLASSICAL SPACE

1. Continuity and Individuation

Classical philosophy and the science that grew out of it are rooted in the drive toward individuation. To this end, the prime directive of the classical enterprise is to wrest order from the chaos of nature. As we saw in the Preface, pre-Socratic philosophy viewed nature in the raw as *apeiron*, the Greek word meaning "limitless," "boundless" or "indeterminate." "*Apeiron*" is "the unintelligible; the many; the moving; the ugly; the bad...the inchoate flux of opposites or contraries...the principle of disorder or disharmony" (Angeles, 1981, pp. 14–15). In the proto-scientific practices of alchemy, primordial nature is called *prime matter*—"prima materia, which is the original chaos and the sea" (Jung, 1970, p. 9). The aim of early science and philosophy was to tame and transform "Mother Nature." The emergent individual must extricate himself from her turbulent waters. Out of the chaotic womb of nature, "one" must forge a singular identity; from the many of *apeiron*, must come *one*. A detached posture must therefore be assumed so as to stabilize and fix one's position in the world. In this way, one makes a place for oneself in a world that, for its part, is now transformed. The boundless activity of wild nature is subdued. From irrational *apeiron* comes *peras*—action that is rationalized, subjected to ratio, measurement; action that therefore is limited, held within boundaries, contained within a changeless framework. By thus stabilizing his position in relation to a stabilized world, "man," the detached subject, can exert his influence over nature; he can treat it as an object; parcel it out; measure it with ever greater precision. For, as Protagoras said, "man is the measure of all things."

Classical thought comes into its own in the writings of Plato. In the *Timeaus*, Plato states that "we must make a threefold distinction and think of that which becomes, that in which it becomes, and the model which it resembles" (1965, p. 69). The first term refers to any particular object that is discernible through the senses. The "model" for the transitory object is the "eternal object," the changeless form or archetype. This perfect form is *eidos*, a rational idea or ordering principle in the mind of the Demiurge. Using his archetypal thoughts as his blueprints, the Divine Creator fashions an orderly world of particular objects and events. As for "that *in* which [an object] becomes," Plato speaks of the "receptacle," describing the latter as "invisible and formless, all-embracing" (1965, p. 70). It is the vessel used by the Demiurge to *contain* the changing forms without itself changing (1965, p. 69). Plato goes on to characterize the receptacle as *space* (1965, pp. 71–72).

Where does *apeiron* enter Plato's account? When the Demiurge transposes his divine thinking into earthly patterns, the copies he creates of his perfect ideas are themselves *im*perfect. That is because the transposition process requires the Demiurge to work through the *prima materia*, the "primitive chaos" (1965, p. 72) of irrational becoming. Plato puts it as follows:

> There were, before the world came into existence, being, space, and becoming, three distinct realities. The nurse of becoming was characterized by the qualities of water and fire, of earth and air....[B]ut as there was no homogeneity or balance in the forces that filled [space]; no part of it was in equilibrium, but it swayed unevenly under the impact of their motion....[Before] the four basic constituents...were arranged into an ordered universe...they were all without proportion or measure; fire, water, earth, and air bore some traces of their proper nature, but were in the disorganized state to be expected of anything which god has not touched, and his first step when he set about reducing them to order was to give them a definite pattern of shape and number. We must thus assume as a principle in all we say that god brought them to a state of the greatest possible perfection, in which they were not before. (1965, 72–73)

Of course, the "greatest *possible* perfection" is not the same as *total* perfection. What this means is that the Demiurge did not quite succeed in taming wild nature, as Plato readily admitted. Stated in terms of *space*, the receptacle in which nature was to be subdued actually did not perform its function with complete efficiency. Instead of containing the forces of chaos in a manner that kept them fully in check, the container was subject to their disruptive influences, and this was true even *after* the formative hand of god was felt. The philosopher of science John C. Graves (1971), in his description of the inhomogeneity of Platonic space, notes accordingly that the "irrational motion," the "fleeting potencies and constantly changing tensions [that filled the receptacle]...persist even after the action of the Demiurge, and thus constantly tend to undo his work" (p. 71). Therefore, while the receptacle was supposed to contain change without itself changing, we can see that this was not entirely the case. With the still potent forces of wild nature forever threatening to break free, and, indeed, periodically succeeding at this, the receptacle was in fact rather given to porosity: at any time, it could "spring a leak." In other words, the inhomogeneity of Platonic space made it susceptible to being ruptured, to losing its continuity.

However, for his part, Graves argues for the complete *continuity* of the receptacle. He claims that it "can have no holes or discontinuities, since this would mean either that it is possible for bodies to move into these holes and thus no longer be in [space], or else the holes would be forbidden to bodies, and there is then no reason why any such regions should exist" (1971, pp. 70–71). A little later, Graves says the following about the original condition of chaos: "In such a state of random, chaotic activity, it is hard to see how regions could be distinguished in such a way as to make a notion of discontinuity meaningful" (p. 71). These conclusions seem to indicate that Graves has fallen prey to the elusive subtlety of the continuity idea.

Evidently, Graves views a "hole" in the continuum as a well-differentiated region that one could enter or leave, or perhaps a region that one is forbidden from entering. In making this assumption, he appears unwittingly to impose the idea of *continuity* upon the "discontinuity" he alludes to. By no means could the absence of continuity be circumscribed within a well-bounded region of space that one could enter or leave, since such boundedness or differentiability itself *presupposes* continuity—as will become clearer in the more detailed study of continuity to follow below. Discontinuity does not require that regions of space be distinguishable, as Graves appears to think, but, on the contrary, precisely that they *not* be. Therefore, the original condition of chaos is one in which it is continuity that is lacking, not discontinuity. And the fact that—despite all the efforts of the Demiurge—*apeiron* is still in evidence in Platonic space, attests to the lingering absence of continuity here. The Platonic continuum is but a *proto*-continuum.

Nevertheless, Plato's receptacle, "leaky" though it was, represented a significant advance in the classical program of containing *apeiron*. And this nascent idea of spatial continuity was brought to fruition with the rebirth and refinement of classical thought at the time of the Renaissance, and in its aftermath. The continuum, in its more mature form, was no longer seen to be buffeted by the irrational forces of prime matter. The guiding influence in ridding space of chaotic matter, thus stabilizing it, was the work of René Descartes. Unlike the Platonic receptacle, "Descartes's space...is perfectly homogeneous" (Graves, 1971, p. 84). Only now—with all the holes in Plato's "leaky vessel" seemingly plugged—does space appear to be completely continuous.

To be sure, when Descartes proposed his new way of thinking about space, he was not speaking about the evolution of space itself but only of our understanding of it. Descartes offered an intuitive clarification of what he believed to be given to us *a priori*, a self-evident universal truth that is to be taken as granted in advance, thus not susceptible to modification. For Descartes, space *always already is* completely continuous, though Plato may have lacked the insight to appreciate this. Descartes related his continuum to the idea of *extension.*

Consider, as an illustration, the simplified space represented by a line segment. In the Cartesian approach, it is intuitively self-evident that the line, however short, has extension. It must then be continuous: it can possess no holes or gaps in it, since, if the point-elements composing it were not densely packed, we would not have a line at all but only a collection of extensionless points. The quality of being extended implies the infinite density of the constituent point-elements.

Yet, at the same time, intuitive reflection discloses the paradox that the absence of gaps in the continuum not only holds this classical space together but also permits it to be *indefinitely divided.* Without a gap in the line to interrupt the process, there is no obstacle to the endless partitioning of it into smaller and smaller segments. As a consequence, though the points constituting this continuum indeed are densely packed, they are distinctly set apart from one another. However closely positioned any two points may be, a differentiating boundary permitting further division of the line always exists (from this we see that a *hole* in the continuum, rather than constituting a well-differentiated region of space, would actually entail a breakdown in spatial differentiability). As Milič Čapek put it in his critique of the classical notion of space, "no matter how minute a spatial interval may be, it must always be an *interval* separating two points, each of which is *external* to the other" (1961, p. 19).

The infinite divisibility of the extensive continuum also implies that its constituent elements themselves are unextended. Consequently, the point-elements of the line can have no internal properties, no structure of their own. An element can have no boundary that would separate an interior region of it from what would lie on the outside; *all* must be "on the outside," as it were. In other words, the Cartesian line consists, not of internally substantial, concretely bounded entities, but only of abstract boundedness as such (Rosen, 1994, p. 92). Sheer externality alone holds sway—what Heidegger called the "'outside-of-one-another' of the multiplicity of points" (1927/1962, p. 481). Moreover, whereas the point-elements of classical space are utterly unextended, when space is taken as a whole, its extension is unlimited, infinite. Although I have used a finite line segment for illustrative purposes, the line, considered as a dimension unto itself, actually would not be bounded in this way. Instead of its extension being terminated after reaching some arbitrary point, in principle, the line would continue indefinitely. This means that the sheer boundedness of the line is evidenced not only locally in respect to the infinitude of boundaries present within its smallest segment; we see it also in the line as a whole inasmuch as its infinite boundedness would be infinitely extended. Of course, this understanding of space is not limited to the line. Classically conceived, a space of any dimension is an infinitely bounded, infinitely extended continuum.

Naturally, it would be a category mistake to interpret the infinitude of classical space as a characteristic of what is *object*. Space is not an object but is the "receptacle" of the objects, the changeless context within which objects are manifested. This distinction, initially made by Plato, is reflected in the thinking of Kant, who held that perceptions of particular objects and events are contingent, always given to variation, but that perceptual awareness is organized in terms of an immutable intuition of space. In the words of B. A. G. Fuller and Sterling McMurrin, Kant took the position that "no matter what our sense-experience was like, it would necessarily be smeared over *space* and drawn out in *time*" (1957, Part 2, p. 220). (I will explicitly address the role of time in due course.) Implied here is the categorial separation of *what* we observe—the circumscribed objects—from the *medium* through which we make our observations. We observe objects *by means of* space; we do not observe space. It is within the infinite boundedness of space that particular boundaries are formed, boundaries that enclose what is concrete and substantial. The concreteness of what appears within boundaries is the particularity of the object. In short, an object most essentially is that which is bound*ed*, whereas space is the contextual bounded*ness* that enables the finite object to appear.

The spatial context is what mediates between object and *subject*. The latter (personified by the Demiurge in the *Timeaus*) is the third term of the classical account and corresponds to what is *un*bounded. That an object possesses boundaries speaks to Descartes's characterization of it as *res extensa*, "an extended thing": what has extension will be bounded. In contrast, the subject is *res cogitans*, a "thinking thing." Entirely without extension in space, the subject has no boundaries or parts. As a consequence, it is indivisible. This is etymologically equivalent to stating that the subject is an *individual*. We may say of the Cartesian individual what we said of Cartesian space: that it is more mature, more fully developed than its Platonic counterpart. To better appreciate Plato's idea of the subject or self, let us consider a still earlier sense of personal identity.

In the mythic epoch that prevailed before the rise of classical Greece, the individual was distinctly lacking in autonomy. According to the cultural philosopher Jean Gebser (1985), mythic experience was not so much actively orchestrated by the directed volition of the conscious individual as it was passively "undergone" (p. 251), "as in a dream" (p. 67). Gebser thus portrayed "dream-like mythical consciousness" (p. 302) as "egoless" and "'we'-oriented" (p. 149). The weakness of mythic individuality was reflected in the weakness of the divinities that were projected by the men and women of the time. By no means did the diffuse pantheon of gods that reigned in these times constitute an omnipotent agency. For, "above the gods is fate, a blind, inscrutable 'will' to which even Zeus must yield" (Jones, 1952, p. 29). The fate that ruled the gods was depicted as a mysterious goddess, or three such womanly figures, *the fates*. Their "blindness" and "inscrutability" speak to their irrational character. According to Francis Cornford (1957), this early notion of fate or necessity is just the opposite of what it has come to mean to us today. Mythic necessity is not that which is "fixed, permanent, unalterable, knowable beforehand," but is "the indeterminate, the inconsistent, the anomalous, that which can be neither understood or predicted" (p. 171). In other words, the "fate" that governed the mythic gods was *apeiron*. Thus, the mythic individual was, in effect, "two steps removed" from full-fledged autonomy: S/he was ruled by the gods, who were ruled, in turn, by the "fickle finger of fate," by the goddess *apeiron* (known also as *Mother Nature*). From this we can see why Gebser (1985) would conclude that mythic consciousness was essentially "spaceless" (p. 117). In myth, the chaotic potencies of *apeiron* went largely uncontained.

The rise of classical Greece constituted a refinement and sublimation of mythic experience. The influence of the old deities began to recede and an enhanced sense of personal identity and agency gradually came to the fore. This heightening of human individuality was manifested in the growth of critical thinking and reflective intellectuality, particularly among those members of society with sufficient leisure time to engage in such activities. A philosopher-class emerged, and, with it, an increased attitude of skepticism toward the ancient gods (among Sophists like Protagoras, for example). Then, with Plato, the process of transforming mythic identity reached its culmination.

Evidenced in the writings of Plato is a curious mixture of incisive critical thinking and a religious faith that might appear to hark back to the age of the myths. Indeed, Plato presupposed the form of identity that constituted the very core of mythical consciousness: the *soul*. ("[T]he essential characteristic of the mythical structure," says Gebser, "is the...*awareness of soul*"; 1985, p. 61.) For Plato, a human being in its essence is not a mortal ego grounded in the material body but a spiritual presence centered in the soul. And yet, Plato's treatment of the soul was hardly a regression to original mythic experience. As is abundantly clear from works such as the *Phaedo* and the *Republic*, Plato sought to imbue the soul with reason. Mythic identity, ruled as it was by *apeiron*, was largely irrational. In the Platonic transformation of the soul, *apeiron* was now to be contained.

I noted above that, in the *Timeaus*, the containment of *apeiron* was the function of the Demiurge. This was the deity who projectively personified the new rational self. The god of Plato was in fact a far more potent figure than had been Zeus and his companions in the pantheon. In the Demiurge, the diffuse powers of the old mythic gods had become concentrated in the power of a singular being, an active agent who—in gaining greater control over *apeiron*, in containing the tempestuous goddess in his receptacle—was now capable of bringing order to all of creation. Of course, the Demiurge had not gained *complete* control over *apeiron*. We have seen that the irrational potencies of *apeiron* were forever threatening to spill out of the spatial vessel, so that an ongoing effort was required of the deity to keep them contained. Yet he did manage to contain them much of the time. What I am proposing here is that the Platonic Demiurge, far from being a reversion to the pre-rational divinity projected by the mythic self, was the prototype of a *new and more individuated self*. Following Plato's time, this nascent identity would be projected as the even more individuated monotheistic god of Judeo-Christianity, and, after incubating in that guise for over a millennium, it would begin to come into its own at the time of the Renaissance, when the projection of the Judeo-Christian *deus ex machina* began to be withdrawn.

With the rise of Judeo-Christian monotheism, a whole new experience of divinity came to prominence. The Hebraic Jahweh was both a more powerful God than was the Demiurge, and a more personal one. For Jahweh, there is no ongoing struggle to contain *apeiron*, since, in effect, it is He who creates chaotic *apeiron* to begin with, then quickly subdues her. We see this in the opening passages of the Old Testament:

> In the beginning God created the heaven and the earth.
> And the earth was without form, and void [it was *apeiron*]; and darkness *was* upon the face of the deep. And the Spirit of God moved upon the face of the waters.
> And God said, Let there be light: and there was light.
> And God saw the light, that it was good: and God divided the light from the darkness…

It follows from this that the Judaic cosmos is more homogeneous than was Platonic proto-space. The formless energies of dark nature are created *ex nihilo* by a rational God, who moves at once to contain them, and has little difficulty in doing so. "Let there be light," says He. And no sooner was this said than it was done.

Concomitant to the Judaic enhancement of divine power was its internalization. There was now a *personal covenant* with God through which the capacities of the human individual were themselves expanded. By faithfully upholding the sacred pact, by following God's commandments from day to day and moment to moment, the individual was better able to order his or her life, to gain a greater measure of control over it. Thus, the new, intensified, more intimate relationship with divinity marked a further crystallization of human individuality, with less projection of effective agency to an external source.

Judaic individuation was brought to a culmination with the emergence of Christianity. For, with the coming of the Messiah, there was not merely a close personal relationship with the divine; rather, God had come down to earth, had become incarnated in the form of a human being. As Richard Tarnas puts it, "In Christ...the historical and the timeless, the absolute and the personal, the human and the divine became one" (1991, p. 102). Nevertheless, despite the symbolic union of humanity and God through the mediation of Jesus, the distinction surely had not simply been abolished. Yes, Christ was somehow both divine and mortal. He had appeared on earth, gone to the cross, was resurrected, then reunited with His Father in heaven. The promise was that Christ would come again to redeem humanity, to lead it into the kingdom of heaven. But, in the meanwhile, a great chasm still existed between the power and glory of the Creator, on the one hand, and the earthbound mortals he had created, on the other. It was true that, by surrendering himself to Jesus, a human being could align himself with God. So close could the person of faith draw to God—yet so far away God remained.

The disintegration of classical civilization was completed around 500 C.E. In the millennium that followed, the Church reigned supreme. During the Middle Ages, there was little overt evidence that individuation was being further advanced. Then, after this period of gestation, a whole new order of identity began to emerge. By no means was the Re-naissance merely a rebirth of the post-mythic individual of classical Greece. Indicated now was the appearance of a new post-*Christian* individual. Two thousand years after the old mythic sense of identity was transformed by the Greeks, Judeo-Christian individuality started undergoing its own transformation.

According to Gebser (1985), with the Renaissance, "Man is henceforth not just in the world but begins to possess it; no longer possessed by heaven, he becomes a conscious possessor—if not of the heavens, at least of the earth" (p. 12). Gebser dramatizes this transformative secularization—and its inseparability from a new awareness of *space*—by citing a 1336 letter of Francesco Petrarch, the discoverer of landscape. Petrarch describes his ascent of Mount Ventoux, an experience that "unsettles" him and produces "intense agitation": "'Shaken by the unaccustomed wind and the wide, freely shifting vistas, I was immediately awe-struck'" (cited in Gebser, 1985, p. 14). Petrarch's letter is written as a confession of an epiphany that transforms his life. He goes on to say that "'[W]hat I experienced today will surely benefit myself as well as *many others*'" (p. 14). Gebser comments that the experience in question was nothing less than an "encounter with what was then *a new reality*....The event that Petrarch describes in almost prophetic terms as 'certainly of benefit to himself and many others' inaugurates a new realistic, individualistic, and rational understanding of nature" (pp. 14–15). The trip to the summit of Mount Ventoux and subsequent inspection of the vast panorama that lay below is portrayed by Gebser as the first movement into a "free, open, and unenclosed space" (p. 15), an "elemental irruption of the third dimension and transformation of Euclidean plane surfaces" (p. 16).

The new space to which Gebser alludes far surpasses the "leaky vessel" of the proto-Christian Demiurge and even the "better insulated" Judeo-Christian cosmos. With the Renaissance, the continuum now manifests itself as *thoroughly* sealed, entirely homogeneous, with *apeiron* nowhere to be found. Evidently, it is no longer the case that order must be brought to space by the divine act of subduing *apeiron*. Instead, space is considered to be well ordered *from the first*, its continuity being taken *a priori*.

Whereas Renaissance space is constituted by sheer extensiveness and the objects extended therein are fully constrained by its laws (the laws of motion), the Renaissance subject is utterly *un*extended, thus freely transcendent of space. This speaks to the self-possessed autonomy of the new individual: more decisively than ever before, the individual is "his own man." In freeing himself from heaven, from the erstwhile constraints of God and religion, he does not simply return to the earth. On the contrary, he stands apart from that as well. For, as Gebser suggests, the Renaissance man *possesses* the earth, which means that, instead of merely becoming immersed in worldly matters, he turns them into his objects so he can shape them to his will. Thus, in standing apart from God, the newly secularized subject himself becomes god-like: He possesses without being possessed, and what he possesses, what he shapes and molds to his desires, (re)creates in his own image, are the objects cast before him in space (the word "object" comes from the Latin, *objicere*, "to cast before").

The Renaissance subject or "I" can be said to function as an "*eye*": it is a point of view, a window or perspective upon the objects. Accordingly, Gebser designates the new world of the Renaissance as the "*perspectival* world." The development of the technique of perspective in art indeed was a defining innovation of the Renaissance and Gebser goes to pains to describe its historical unfoldment during the fourteenth to sixteenth centuries (1985, pp. 11–20). In the course of his account, Gebser notes "perspectival man's predominantly visual or sight orientation" (p. 18), and points out:

> The basis of the perspectival world view is the visual pyramid; the two lines extend from the eyes and meet at the object viewed. The image formed...includes the subject, the object, and the space in between. Piero della Francesca clearly expresses this in his remark: "The first is the eye that sees; the second, the object seen; the third, the distance between the one and the other." (Gebser, pp. 18–19)

Citing Erwin Panofsky's observation that "'perspective provides a distance between man and objects,'" Gebser further comments:

> Such detachment is...a sign of an emergent objectifying consciousness....The emergent awareness of distantiating space presupposes a clear vision; and this heightening of awareness is accompanied by an increase of personal ego-consciousness. (p. 19)

The Renaissance differentiation of "the subject, the object, and the space in between," clearly parallels the Platonic "threefold distinction...of that which becomes, that in which it becomes, and the model which it resembles." Both formulations reflect the crux of classical cognition, the axiomatic base serving as its unquestioned point of departure—the self-evident intuition of *object-in-space-before-subject*. The object is *what* is experienced, the subject is the transcendent perspective *from* which the experience is had, and space is the continuous medium *through* which the experience occurs. The relationship among these three terms is that of categorial separation.

Of course, the *Platonic* version of this trichotomy is less well developed than its Renaissance successor. Again, Plato's "model" is *eidos*, a formative idea in the mind of the Demiurge, who is the divine subject seeking to impose order upon matter. Because this subject is still struggling with *dis*order, with *apeiron*, it is a less effective agent than the transcendent subject of the Renaissance. The Platonic subject is less able to stand apart, to gain a detached perspective upon nature. It therefore cannot fully *contain* the irrational potencies of nature. So, "that *in which* [nature] becomes," Plato's "receptacle" or space, is a "leaky vessel." Spatial continuity will be realized more completely only when Platonic subjectivity is superseded and a new step taken toward individuation. It is for the *Renaissance* subject that nature appears fully contained, securely enclosed within the "uniform space-time context." The latter was Heidegger's term (1962/1977, p. 268) for the Galilean framework in which all bodies could be precisely located, probed and experimented upon, submitted to mathematical analysis, and, ultimately, shaped and manipulated through the enterprises of science and technology. The crystallization and consolidation of space thus brought enhanced "objectivity," an unprecedented capacity for one to focus clearly and sharply on what lay before one. The objects of nature were now laid open, fully exposed, extended for close scrutiny (they were *res extensa*), whereas the subject—being utterly *un*extended in Descartes's new definition of mind or spirit—was closed to view, essentially anonymous. Consciousness thus became "irreversible." With the Renaissance, the subject—having become wholly detached and disembodied, thus rendered anonymous—was not to *be* known; *it* was to do the knowing.

But *was* the process of individuation really completed with the Renaissance? Was the subject now *fully* transcendent of space and the objects contained therein? Was space now *completely* continuous, a perfectly homogeneous context in which the Cartesian agent could operate with impunity to gain mastery over nature? To put the question a little differently, was *apeiron* entirely eliminated in the Renaissance experience of *object-in-space-before-subject*? We are going to see that, in all cases, the answer is no.

2. The Challenge to Continuity

A. Introduction: The Mechanical Age and the Ideal of Continuity

The space of the Renaissance is mapped by the system of coordinates Descartes invented for the purpose. At the center of this X-Y coordinate system is the 0,0 origin (Fig. 1).

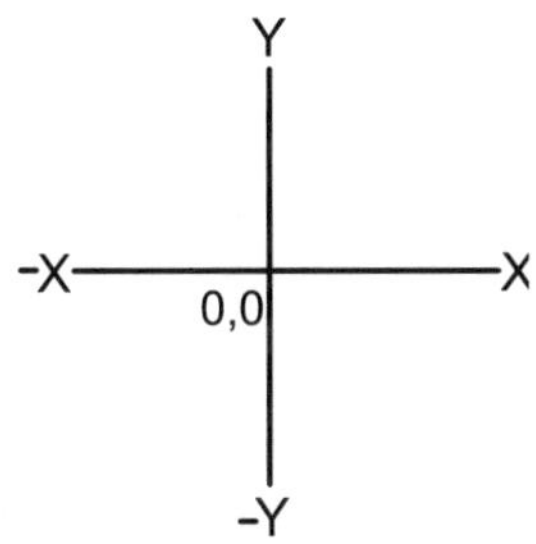

Figure 1. Cartesian coordinates

The origin of the reference frame plays a unique role. Being the locus at which observation begins, 0,0 serves, in effect, as the surrogate for the subject's "eyes," as that from which a perspective upon the world opens out. To be sure, each time I change my point of view, my position in space changes and the 0,0 origin shifts accordingly. However, these transformations of the coordinate system are commensurable. They are encompassed in a single system of measurement, so that I can pass continuously from one frame of reference to the other. In this way, change in perspective is contained, subsumed within the changeless continuity that is space. Because one perspective can readily be transformed into the other by a continuous displacement in space, perspectives can be said to be equivalent.

The equivalence of perspectives in the new Cartesian space was dependent on the emergence of the new order of subjectivity, Descartes's *cogito* or *thinking* subject (what Heidegger called the "special subject"; see 1962/1977, p. 273). The *cogito* was a more abstract individual than was his pre-Renaissance predecessor; by detaching himself more completely from the earthly realm, he could transcend all particular standpoints and view reality from a universal frame of reference. It was from that point of view that particular perspectives were rendered equivalent. Concretely changing viewpoints are commensurable because they are undergirded by the changeless universal perspective of the *cogito*. Thus, whereas each 0,0-centered coordinate system is specifically situated in space, the generic "I" that lies behind these specific "eyes" has no such fixed position in the extensive continuum. As noted above, Descartes's *cogito* is *un*extended, transcendent of space.

Of course, the *cogito* is a theoretical entity. In the practical operations of post-Renaissance science, the presence in space of a human observer was required to make measurements and conduct experiments. Nevertheless, it was possible to transcend theoretically the observer's local perspective, his or her local space and time, and to assume the universal outlook, that from which space and time are absolutely invariant. Thus, in this classical idealization, the standpoint of the actual observer was superseded by the pan-optical, "totally objective" viewpoint of a god-like observer.

In the language of classical physics, the perspective (reference frame, coordinate system) of the ideal observer is "inertial." An inertial frame of reference is one in which the laws of classical mechanics have full validity. In particular, inertial systems are subject to Isaac Newton's law of inertia, which states that a body at rest or in uniform motion will remain at rest or in uniform motion unless some outside force is applied to it. If we consider two such inertial systems, it is impossible to determine their respective states of motion independently of each other. Since the motion of one can only be established in relation to the other, we cannot say in absolute terms that a given system is at rest, or that it is moving. For example, as I sit here at my computer, I seem to be at rest, but, from the frame of reference of an astronaut, I am moving with the earth. I may want to know whether I am moving or I am not. But there is no absolute answer; it all depends on the particular reference frame that is adopted. This is the principle of Galilean relativity.

Nevertheless, on this classical view, there clearly is something that is *not* relative, something that does not change in going from one coordinate system to another: the laws of classical mechanics. Assuming that the astronaut and I are moving uniformly with respect to one another, the Newtonian laws governing the behavior of bodies in space and time are equally applicable, regardless of which frame of reference is chosen. We are therefore able to perform a Galilean transformation. My state of motion relative to my own reference frame (that wherein I seem to be at rest) may be converted into my state of motion relative to the astronaut's coordinate system (wherein I appear to be moving with the earth). These inertial reference frames are entirely interchangeable. They are but particular variations on the universal viewpoint that is itself invariant. From this point of view, it is not so much concrete objects that are observed; rather, what are observed are the changeless laws of classical mechanics. This is the idealized, "purely objective" reference frame of the generic Cartesian subject, that from which we can indeed speak of absolute rest, perfect inertia. Thus, whereas particular perspectives or coordinate systems are completely relative, the more abstract frame of reference of the *cogito* is not.

In the foregoing classical account, we recognize the formula that derives from Plato: *object-in-space-before-subject.* Plato's intention was that the only term of the formula that would be susceptible to change would be the *object.* Both the subject (the Demiurge) and the spatial continuum in which he operates were to be wholly invariant. But Plato did have his problems with *apeiron*, as we have seen. Apparently, it was Descartes and Newton who finally succeeded in banishing *apeiron*, and thereby fully stabilizing the subject and its sphere of action once and for all. Or did they?

B. Doubts about Continuity in the Age of Light

The complete banishment of *apeiron* requires complete agreement between theoretical and empirical reality. For the chaotic divergences of the material realm to be reined in, for unitary order to be fully imposed, local observations must be made fully consistent with the *cogito's* global perspective. Was this actually achieved in the physics of Galileo, Descartes, and Newton? In a paper titled "Global Idealism/Local Materialism," Koichiro Matsuno and S. N. Salthe (1995) point out that the theoretical assumption made by classical mechanics of a "Laplacean demon," a global "observer with an instantaneous bird's eye view of everything" (p. 311), was not much more than an article of metaphysical faith—one that had not been definitively reconciled with the "worm's eye" view of the empirical observer gathering actual data down on the ground. Or, as Gary Zukav (1979) put it in his popular treatment of the history of physics, "No one [had actually] found a co-ordinate system in which the laws of mechanics [were] valid!" (p. 148). Therefore, while physics had *assumed* the existence of an ideal inertial system, one that was perfectly at rest, the truth was that no such system had been unequivocally demonstrated; measurements always revealed small discrepancies between theoretical expectations and values that were actually obtained. Though future refinements in measuring devices might eventually bring theory and data into perfect alignment, in the absence of definite proof there remained the possibility that concrete differences in perspective might turn out *not* to be wholly equivalent when regarded from the abstract point of view; that Galilean transformations from one coordinate system to another might *not* be perfectly continuous; that the extensive continuum itself might be something of a fiction. What all this intimated was the possibility that *apeiron* was still at play!

Nevertheless, there was no disputing the success of post-Renaissance physics in transforming human culture. As a result of the early efforts of Galileo, Descartes, and Newton, science and technology came into their own, and this, in turn, led to the creation of a whole new world. It stood to reason then that, while the *perfection* of the generic Cartesian subject might have been fictitious, this abstract individual was real enough, at least in terms of his practical impact. Moreover, in the age of classical mechanics (sixteenth to nineteenth centuries), the idealization of the *cogito* was clearly tolerable, since it was possible effectively to ignore limitations on the precision with which the motions of bodies could be measured. Approximation could be taken for reality with no major consequences to the scientific enterprise. Then, in the nineteenth century, came the *Age of Light*.

In the Preface to this book, I cited Jaspers's (1941/1975) comment on how science changed in the century that followed the death of Kant: "It extends further than in Kant's time; it is more radical than ever, both in the precision of its methods and in its consequences" (1941/1975, p. 166). But Jaspers goes on to tell us that, in the very extension and refinement of science, the *limitations* of scientific knowing have become much more transparent: "We experience the limits of science as the limits of our ability to know and as limits of our realization of the world through knowledge...the knowledge of science fails in the face of all ultimate questions" (1941/1975, p. 167). For my part, I would emphasize that the barrier science reached as it progressed into the nineteenth century was not merely an external one. It was not simply that the scientific approach was found to be inapplicable to traditionally nonscientific fields of knowledge. It was that fundamental problems arose within science itself. As physics pressed toward ever higher levels of exactitude, extending itself to the extremes of measurement, to the limits of its scales—ultra-high velocities, sub-microscopic distances, etc.—some of its most basic expectations were upset. The initial upheaval came with the failure of Galilean relativity implicit in the experiment on light done by Albert

Michelson and Edward Morley in 1887. Let us consider the historical background of this momentous occasion.

In the course of the nineteenth century, concern with the influence of relatively crude mechanical forces upon ponderable bodies gave way to an in-depth investigation of the more subtle phenomena of electricity and magnetism. By the 1870s, the research of Michael Faraday and James Clerk Maxwell had culminated in the theory that all forms of electromagnetic energy, light included, are propagated as waves. It seemed obvious that, even though an electromagnetic wave was a more elusive phenomenon than were mechanical waves such as those found in water or air, it nevertheless required a *medium* through which it was transmitted. Without such a medium, it would be difficult to see how the energy could be propagated in a continuous, objectively observable fashion. Just as a water wave travels through water and a sound wave travels continuously through air, a light wave must have its own, more subtle medium through which to travel. The implication of this attitude was that electromagnetism was still essentially to be viewed in mechanical terms, as a classical phenomenon—as an *object in space*.

The medium posited for the transmission of electromagnetic waves was the "luminiferous ether." The ether was said to pervade the entire universe, and was alleged to be a paragon of purity. So fine-grained and subtle was the "ether sea" that it was utterly impossible to detect. Possessing few properties and no action of its own, it served exclusively as the framework within which the motions of coarser substances could be measured and analyzed mechanically—including the motion of light. Being completely immobile, absolutely at rest, the ether would be the decisive inertial system; it would provide spatiotemporal continuity *par excellence*. It would be the coordinate system that would anchor the transformation of all other coordinate systems, rendering them all completely equivalent in the universal view of the transcendent *cogito*. This had been the underlying hope and expectation of classical physics all along and it would now be confirmed. Plato's "leaky old vessel" would be sealed hermetically once and for all. *Apeiron* would at last be banished in a definitive way. I suggest that, at bottom, it was not just the success of physics that was at stake in verifying the existence of the ether. Well below the threshold of what nineteenth century physicists were prepared to acknowledge, something more basic was at risk. The stable sense of individual identity that had been cultivated from at least as far back as the mythic epoch was now opened to question, as we will shortly see.

An important milestone of mid-nineteenth century science that made it possible to test the theory of the ether was the development of the ability to directly measure the velocity of light, achieved by Armand Fizeau in 1849. It was this that Michelson and Morley depended upon in devising the interferometer for their crucial experiment. If it was true that light moved through the motionless ethereal continuum, then a key principle of Galilean transformation should apply: the addition of velocities. Assuming light to propagate through the ether at the absolute speed of *c* (approximately 186,000 miles per second), a traveler moving toward a beam of light should observe the beam to be approaching her at a velocity greater than *c*, her own velocity being added to *c* to obtain the higher relative velocity. Similarly, if the light beam and the observer are moving away from each other through the ether, the relative velocity of the light beam would be less than *c*, the observer's velocity now being subtracted from *c*. What Michelson and Morley discovered was that the velocity of light actually always appeared to be the same, regardless of its direction of motion relative to the observer. This astonishing result sounded the death knell of the ether theory. Let us consider it in more detail.

The failure of Galilean relativity was indeed baffling to the classical "eye." Is it not an obvious fact of perception that, if I change my perspective on an object I am viewing, its appearance will change accordingly? What the Michelson-Morley experiment demonstrated in its abstract way was that, when the "object" being considered is *light*, the familiar principle of experience does not apply. It would certainly look strange to me if I got up from this computer screen I am sitting at, moved all the way to the right of it so that I was viewing the screen at an acute angle, but found that the screen had the same full, square appearance as when I was sitting directly in front of it! Analogously, this is what Michelson and Morley "saw" when they looked at light from different "angles" (reference frames). This strange outcome was most alarming to physicists. It came as a bombshell that was soon to precipitate a revolution in physics (the Einsteinian revolution; see below). That is because the finding of Michelson and Morley did nothing less than call into question the classical intuition of object-in-space-before-subject that had implicitly governed human experience for many centuries.

We know that the maintenance of classical identity is essentially predicated on separating subject and object in such a way that, whereas objects change concretely, the subject, and the space in which it operates, remain abstractly the same. The concrete change I expect to see when I shift my position with respect to the computer screen presupposes that the spatial context within which I make this observation itself does *not* change, that its continuity is preserved. By the same token, the presupposition is that *I* do not change, that the "invisible I" behind these eyes remains in-*di*-visible, retains its in-dividuality. It is because a solitary individual lies behind the multiplicity of concretely shifting perspectives that said perspectives are readily interchangeable, with one being continuously transformable into another in a predictable fashion. Confirmation of these predictions serves to confirm the *cogito's* sense of himself. When, upon altering my perspective, I see how the appearance of an object I have been viewing changes as expected, I am indeed reassured that I have *not* changed. What Michelson and Morley saw was not reassuring.

To appreciate just how deeply troubling the Michelson-Morley experiment was, let us digress briefly to return to the root of the problem. Cultural theorist Elizabeth Grosz (1994) examines the original condition for the emergence of individuality:

> For the subject to take up a position as a subject, it must be able to be situated in the space occupied by its body. This anchoring of subjectivity in its body is the condition of coherent identity, and, moreover, the condition under which the subject *has a perspective* on the world, and becomes a source for vision, a point from which vision emanates and to which light is focused. (p. 47)

The emergent subject is "the focal point organizing space. The representation of space is...a correlate of one's ability to locate oneself as the point of reference of space: the space represented is a complement of the kind of subject who occupies it" (p. 47). Fully developed, the space in question is "the perspectival space that has dominated perception at least since the Renaissance" (p. 48); it is the continuum. In sum: "A stabilized body image,...a consistent and abiding sense of self and bodily boundaries, requires and entails understanding one's position vis-à-vis others, one's place at the apex or organizing point in the perception of space" (p. 48). So our earliest, most basic sense of ourselves as stable autonomous individuals is inseparably linked to our experience with space. (The idealized Cartesian subject might fancy itself a disembodied spirit that is freely transcendent of space, but the underlying reality is that—however high this subject may appear to fly—it remains "tethered to its ground." The classical subject's relation to space necessarily is maintained, as the old formula implies: object-in-space-before-subject.)

Grosz's analysis of Sigmund Freud and Jacques Lacan demonstrates the close parallel between the development of subjectivity in the history of human culture, on the one hand, and the early development of an individual child, on the other. Freud's (1914) concept of "primary narcissism" implies that there is a primary constancy grounding all others: the image of one's body, which constitutes the earliest manifestation of the ego. Because one's imagined body serves as the frame of reference from which all observations are made, an object seen to change with changes in perspective can be taken as the same object only insofar as the observer's body is implicitly sensed as remaining the same. Object constancy thus depends on subject constancy, or, in Freudian terms, on the constancy of the ego.

In Lacan's (1953) elaboration on Freud, the "mirror stage" is crucial to the initial appearance of an invariant body image. The child begins to develop a stable sense of its identity, to form an image of its own body as a whole, when the mother's body can be mirrored back to it as its primary object or alter-ego. This possibility arises at around six months of age and depends, in turn, on the psychological separation of the infantile body from that of the mother. Grosz (1994), in drawing on Lacan, thus observes that it is in this mirror stage that "the division between subject and object (even the subject's capacity to take itself as an object) becomes possible for the first time" (p. 32). And, of course, this is the stage in which *space* first appears. It is through space that the confirmation of one's basal identity is mirrored back to one. In encountering expected variations in the appearance of the mother's face with changes in perspective upon her, the emergent individual is reassured that he is indeed *one*. It is this primary reassurance that anchors human identity and eventually makes possible the quest for unity in science.

For many centuries the work of science was most reassuring and ultimate unification seemed attainable. Then came the Age of Light and the crucial experiment of Michelson and Morley. Just why was it that the velocity of light did not change regardless of the frame of reference these researchers adopted? Why did light "look" the same to them no matter what perspective upon it they assumed? I propose it was because, in confronting the phenomenon of light, they were not encountering an object to be seen, but that *by which* they saw. As the physicist Mendel Sachs put it in his inquiry into the meaning of light: "What is 'it' that propagates from an emitter of light, such as the sun, to an absorber of light, such as one's eye? Is 'it' truly a thing on its own, or is it a manifestation of the coupling of an emitter to an absorber?" (1999, p. 14). Sachs's rhetorical question suggests that the phenomenon of light—instead of lending itself to being treated as an object open to the scrutiny of a subject that stands apart from it—must be understood as entailing the *inseparable blending of subject and object.* This computer screen might not look the same to me from every perspective, but would not my *viewing* of the screen look the same? In attempting to observe the light by which the screen is perceived, it seems I would be confronted with the prospect of "viewing my own viewing," and this would mean that I would not encounter the concrete variations in appearance that attend the observation of an object from a viewpoint that itself is not viewed.

Ever since the "mirror stage," what we have viewed has been limited to the object-in-space. Illuminated by light, the myriad objects have mirrored back to us their implicit confirmation of the unity of this subject, this unviewed viewer. But what light *per se* mirrors back is our own *viewing process*, that in which subject and object, viewer and viewed, are indissolubly fused. If we gaze into the "mirror of light" with the old expectations, we are certainly in for a shock. For what we see in the "glass" is *nothing*—that is, no *thing*, no well-defined object of any kind. Light mirrors back to us the "indeterminate," the "unintelligible…[an] inchoate flux of opposites or contraries" (Angeles, 1981, pp. 14–15). It gives us a glimpse of *apeiron*.

If there is no separation of subject and object in the phenomenon of light, surely there can be no *space*, since the existence of space presupposes that separation. The classical formula, object-in-space-before-subject, is hyphenated to indicate that its three main terms are mutually interdependent. Space is therefore no free-standing abstraction. Instead of simply existing on its own, it exists as a *container of concrete objects*, and these objects necessarily are contained in such a way that they are categorically divided from the subject, he who is *un*contained. If the objects were not thus sealed into space, if the subject was not sealed out, the spatial seal would be broken. Just such a breach is signified by the fusion of subject and object encountered in the phenomenon of light. It was for this reason that the findings of Michelson and Morley, instead of confirming the existence of the "ethereal continuum" as expected, pointed to an alarming *loss* of continuity.

The merging of subject and object and concomitant failure of classical continuity clearly was not a welcome sight in the eyes of nineteenth century science. It had been the *separation* of subject and object that had marked the first step toward individuation thousands of years before. In the "mirror stage," the nascent subject had made its initial move toward autonomy by extricating itself from the chaotic womb of nature. In so doing, "wild Mother Nature" had been subdued, transformed into an object-in-space. The projection of object-in-space-before-subject thus constituted the essential act whereby the *cogito* emerged from the irrational flux of *apeiron*. Now, in the nineteenth century, in confronting the phenomenon of light, the projection upon which human identity essentially had come to depend appeared in danger of collapse. To the classical eye, this prospect was truly shocking—nightmarish, in fact. The underlying threat was that of dissolution, regression to an undifferentiated, infantile state of unreason involving nothing less than the extinction of the *cogito*. I do not mean to suggest that these ominous implications were widely or openly discussed, or even consciously recognized by the physicists of the nineteenth century. To most, while the Michelson-Morley finding certainly posed a fundamental challenge to science, it was one that could be addressed in a "scientific" fashion, though an unprecedented level of ingenuity would have to be brought to bear on the problem. This is what the late nineteenth century revolution in physics was all about. In actuality, the "revolution" had a wholly conservative aim. At bottom, what Albert Einstein was seeking to do was *maintain* the classical posture of object-in-space-before-subject.

C. The Unsuccessful Effort to Deny Discontinuity

Einstein's theory of relativity is essentially co-optative. It accommodates the electromagnetic challenge to classical intuition in such a way that the challenge loses its force. The brilliance of Einstein lay in the fact that, at the very same time that he accepted the gap in the classical continuum (he could hardly do otherwise, given the demise of the ether), he found a way to seamlessly fill it in. It is this that is responsible for the popular confusion over the meaning of the term "relative" in Einstein's theory.

Space and time are indeed relativized in Einstein's theory. We know that, prior to Einstein, these dimensions were assumed to be absolute parameters of change that did not themselves change. The space and time of Newton and Descartes were taken as perfectly homogeneous. Unlike Plato's porous receptacle, they were supposedly exempted from the turbulent potencies of *apeiron* that, if violent enough, could wind up rupturing the continuum. What did change on the classical view, what *was* subject to discontinuity, were not space and time themselves but the concrete objects contained therein. How did Einstein respond when the Michelson-Morley experiment cast doubt on the older assumptions about space and time? In effect, he said: Let space and time themselves change (in the theory, time dilates and space contracts at high relative velocities). This transformation of space and time certainly appears to dynamize physics and render it concrete, for now, not only do objective events entail dynamic processes; also in process is what previously had been taken as the abstract, utterly static framework for those events. In thus introducing change at the fundamental level of space and time themselves, Einstein did seem to be challenging the classical order of object-in-space(-and-time)-before-subject. And yet, the philosopher Bertrand Russell (1925) was prompted to declare that Einstein's theory was *misnamed*, since the theory actually seeks a description of nature that is anything but relative!

It is clearly an oversimplification of Einsteinian relativity to say, without qualification, that, in it, space and time change. Einstein did not simply posit the variability of space and time. Instead he declared that, while these terms previously had been taken as invariant, they must now be seen as undergoing change *within a new context of invariance.* Yes, time is now deemed relative. No longer can we assume a universal clock that enables observers in different frames of reference, traveling at different velocities, to synchronize their watches with absolute precision. But while *time* is no longer absolute, *space-time is*. All concrete observational perspectives, involving variations in velocity, are rendered strictly equivalent in relation to the four-dimensional space-time continuum whose unity is conferred by *c*, the constant velocity of light. The Michelson-Morley experiment had intimated the possibility that light's constancy was indicative of a blending of subject and object that confounded classical intuition. Einstein foreclosed this interpretive option before it could even reach the threshold of conscious consideration. For Einstein, light is hardly a merging of subject and object but is simply an abstract object: it is the empirical constant necessary for the objective determination of space-time events. Thus, most essentially, Einsteinian "relativity" was actually not about relativizing or dynamizing nature; it did not embody a genuine recognition that there is fundamental change or discontinuity in the world, that the world is in process. Einstein's success, his profound influence on twentieth century physics, was rooted in his ability to accommodate the nineteenth century challenge to classical physics in such a way that the classical viewpoint is basically upheld. The old order of space and time is supplanted by Einstein, yet, with scarcely a pause, it is replaced by an even more abstract order of this kind: that of the four-dimensional space-time continuum. Here there is still the object, or rather, the objectified relativistic event; still the static continuum that contains the event, divesting it of its vitality; and still the detached subject who analyzes all this from afar. To be sure, Einstein significantly updated the details of the classical formula, but he did this in order to maintain the viability of its basic terms. Let us now examine Einsteinian space-time a little more closely.

In Einstein's special theory of relativity, space and time are united by replacing the hitherto separate measures of distance and duration with a new fundamental unit, the "space-time interval." A constraint placed on this unification is the need to distinguish unambiguously space-time intervals that are "space-like" from those that are "time-like." To put it roughly, although every space-time interval indeed combines space and time, any given interval must be "weighted" more by one than by the other. Essentially, two events in Einsteinian space-time are separated in a space-like way if the distance between them is great enough relative to their temporal disjunction that an observer could be present for the onset of both events only by traveling from one to the other at the velocity of light or greater. In Einstein's theory, no material body can travel that quickly. Events are separated by a time-like interval if the time elapsing between them is long enough relative to their spatial separation that an observer may be present for the onset of both events, but could not experience them as occurring simultaneously, since to do so again would require traveling at the prohibited velocity.

Figure 2 shows the determination of spatiotemporal relations in Einstein's theory. This representation of space-time is called an Einstein-Minkowski diagram, since Einstein drew from the work of the mathematician Herman Minkowski in developing it.

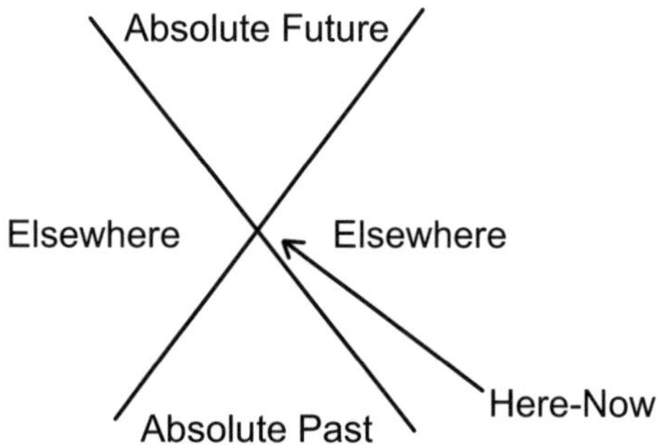

Figure 2. Light cones

Incorporating *c* into the theory results in the formation of a pair of "light cones" through which time-like and space-like space-time intervals are unequivocally established. The structure of the cones thus establishes a clear-cut dichotomy between events that are successive, and those that are contemporaneous. With respect to the temporal aspect, the cone of the absolute future widens upward from the point of the immediate present ("now"), while the absolute past is extended below the now-point. All events in the future cone are definitely *after* the now, whereas those in the cone of the past are definitely *before* it. This is because the time intervals in question are long enough relative to the spatial separation of events that an observer present for the onset of a given event could indeed be present for the onset of a subsequent or antecedent event, but could not experience either of the latter as occurring *contemporaneously* with the given event, since to do so would require traveling at the prohibited superluminal velocity. In these time-like relations, it is clearly impossible for the future and the past to make their presence felt in the now. Regarding the spatial aspect, the light cones intersect at the point of the immediate *here*. Events outside the cones are definitely elsewhere, as opposed to being here, since the space intervals in question are large enough relative to the temporal separation of events that an observer could be present for the onset of both events only by traveling at the forbidden velocity. In these space-like relations, elsewhere cannot be present here.

Like the 0,0 origin of Cartesian space, the here-now originating point of Einstein-Minkowski space-time plays a pivotal role in the classical intuition of object-in-space-before-subject. For, as the locus at which observation begins, the origin functions as the surrogate for the subject's "eyes," its point of perspective upon the world, its "window" upon the objects-in-space. The observer's perspective "centers the light cone of a Minkowski diagram, as the point event at its center: (x,t)=(0,0)" (Schumacher, 1989, p. 111). What is the *difference* between Cartesian and Einsteinian observational perspectives?

We have seen that the older science could ignore the observer's local perspective, its local space and time, and assume a global perspective in which the 0,0 origin is implicitly taken as the "totally objective" viewpoint of a universal observer (a "Laplacean demon"). What Einstein discovered is that, when it comes to the phenomena of contemporary physics—phenomena whose velocities approach the velocity of light—the local space and time of the concrete observer could not merely be discounted. Of course, Einstein did not simply *accept* the intrinsic variability of concrete observation. In his revision of classical physics, concrete observation itself is explicitly included in the account of nature by making this subjectivity into a *new object*, one whose transformations are invariant in a higher-dimensional context. Thus, there is a substantial difference between the pre-Einsteinian and Einsteinian versions of the classical posture. In the former, we have objective events occurring in three-dimensional space before the observing gaze of an idealized subject. In the latter, where subjectivity itself is taken as object, our "objects" are *observational* events transpiring in four-dimensional space-time. Whereas three-dimensional events are concretely observable, the fourth dimension of Einsteinian relativity is an abstraction. The higher-order Einsteinian observer of these four-dimensional acts of observation is therefore a further step removed from concrete reality than was his Cartesian predecessor. Nevertheless, in both cases, the traditional stance is strictly maintained. In both, we have object-in-space-before-subject.

To all appearances, Einstein's special theory of relativity was a resounding success. It had confronted the mid-nineteenth century threat of discontinuity by reestablishing continuity at a higher level of abstraction, that of the four-dimensional space-time continuum. Was *apeiron* finally expelled with this? Had the unitary subject at last gained full mastery over the disconcerting variances of "Mother Nature"? Was the classical project of individuation now realized in earnest?

When Einstein unveiled his special theory of relativity in 1905, he was well aware that it was incomplete. The initial theory was "special" in the sense that it was limited to the ideal case of coordinate systems that moved uniformly. In the real world, however, systems typically *change* their state of motion, speed up or slow down. With the special theory published, Einstein turned to the task of accounting for the relative motion of *all* reference frames, whether or not the motion was uniform. This effort eventuated in the 1915 publication of the *general* theory of relativity.

The central idea of general relativity is the *equivalence principle*. Through an ingenious thought experiment, Einstein was able to demonstrate that all coordinate systems are indeed interchangeable, provided that the equivalence of non-uniform motion with *gravitation* is recognized. We are to imagine an elevator whose cable has snapped, so that it is plunging to the ground. As the elevator falls freely, it is accelerating under the influence of the earth's gravitational field. Or so it seems to an observer witnessing the event from an external vantage point. Einstein noted that a different conclusion would be reached by an observer enclosed within the elevator. To her, *inertial* effects would be in evidence: an object that was dropped would appear to float in mid-air at its point of release (since it would be dropping to earth at the same rate as the elevator and the observer herself); an object given a push would move away uniformly, proceeding in a straight line; and so forth. In other words, the conditions produced inside the freely falling elevator would agree with the Newtonian laws of mechanics. The interior of the elevator would constitute an inertial frame of reference, one in which no acceleration was taking place. Naturally, an observer situated outside the elevator would be inclined to believe that his counterpart on the inside had formed a mistaken impression. But, to Einstein, this conclusion itself would be mistaken. The essential point is that, when gravitation is taken into account, the system's state of motion, uniform or not, is entirely relative to the observer's frame of reference. These differing states of motion are therefore interchangeable: an equation can be written for the continuous transformation of one into the other. Einstein's theory gains full generality by demonstrating the equivalence of all non-uniform motion with gravitational influence. In this way, general relativity achieves greater unity than did special relativity: now it is not merely concrete variations in *uniform* states of motion that are rendered abstractly invariant but variations in *all* states of motion.

To analyze the more general form of motion, Einstein required a more complex mathematical tool. For this purpose, he turned away from the geometric approach of Minkowski, which was limited to flat space-time, and adopted the more comprehensive curved geometry of Bernhard Riemann. The flexibility of Riemannian geometry permitted Einstein to gauge the degree of non-uniformity of motion in precise terms by associating it with the degree of curvature in the manifold. Space-time would be without curvature for systems in uniform motion and would become progressively more curved as the acceleration of the reference frame increased. In terms of the equivalence principle, this meant that space-time would be curved by *gravitational* effects: the greater the gravitational mass of a body, the more it would curve the space-time continuum.

Now, while Einstein found it necessary to adopt this approach, he soon realized that it had its limitations. For there were solutions to the field equations of general relativity that predicted *infinite* curvature. That is, if a gravitational body were massive enough, the curvature of space-time would become so great that a singularity would be produced in the continuum. What this meant is that analytic continuity would be lost and the theory would fail! However, for that to happen, the mass density of the gravitational body indeed would have to be enormous. When the general theory was first propounded in 1915, the existence of such astrophysical bodies was taken as purely hypothetical. But, as the twentieth century wore on, the possibility of stellar objects whose masses were sufficient to produce "black holes" in space began to be considered more seriously (such stars are "black" because their gravitational fields are so strong that light cannot escape from them). This led physicist Brandon Carter (1968) to raise explicit doubts about Einstein's theory: Would it be able to survive its prediction of gravitational collapse? By the end of the twentieth century, empirical evidence for black holes had only grown stronger, and, now, as we begin the twenty-first century, the evidence seems almost irrefutable. One might think that, as a consequence, Einstein's theory might have lost much of its influence. Before considering why that is not the case, let me summarize the theory's course of development and reflect on its meaning.

Einsteinian relativity was initially conceived as an attempt to circumvent the "black hole" that was created when Michelson and Morley failed to confirm the existence of the luminiferous ethereal continuum. Einstein would plug the implicit gap in three-dimensional space by postulating a four-dimensional space-time continuum. To generalize this new account to non-uniform motion, he posited the curvature of space-time. What we are seeing, in effect, is that the four-dimensional method used to compensate for the absence of continuity in three-dimensional space winds up re-introducing *dis*continuity. Even though general relativity permits one to establish an invariance between a system in non-uniform motion and one moving uniformly, an invariance that presupposes continuity, the *greater* the non-uniformity, the greater is the curvature of space-time, and the closer one then approaches to the point where invariance breaks down and continuity is lost. So it seems that the moment curved Riemannian geometry was applied to generalize Einstein's remedy for discontinuity, a new order of discontinuity was prefigured! More bluntly stated, Einstein's solution didn't work; at bottom, it did not effectively address the mid-nineteenth century crisis brought on by the resurfacing of *apeiron*.

Then why does Einsteinian relativity remain so influential? Should its inherent discontinuity not have undermined it by now? There is another approach to contemporary physics, one just as influential as Einstein's, that has created an atmosphere in which discontinuity can better be tolerated and its ultimate consequences better denied.

At the close of the nineteenth century, just around the time when physicists were digesting the Michelson-Morley findings on the velocity of light, another ground-breaking experiment on the phenomenon of light was being conducted. Max Planck was investigating blackbody radiation, the emission of electromagnetic energy in a completely absorbent medium, a closed cavity that does not reflect light but soaks it up, then discharges the energy internally. Classical theory faced a difficulty here that was on a par with the problem produced by the Michelson-Morley experiment. If the traditional analysis was correct, energy should be transmitted in a smooth and continuous fashion. Yet this assumption leads to the peculiar prediction that, if a non-reflective body is exposed to intense heat, it should radiate an *infinite amount* of energy—a result that clearly is not borne out by empirical observation. Planck responded to the contradiction by boldly amending the underlying classical assumption. He proposed that light, rather than radiating in a smoothly continuous manner, is transmitted in discrete bundles, *quanta*. This introduction of discontinuity into the theory now brought a remarkable correspondence with empirical data. The new quantum theory could predict laboratory findings to a high degree of accuracy by adding just one parameter, h. This is the constant of proportionality that relates the energy (E) of a quantum of radiation to the frequency (v) of the oscillation that produced it: E = hv. The numerical value of h is 6.63×10^{-34} joule-seconds. The extremely small value of Planck's constant is consistent with the fact that, in the familiar world of large scale happenings, energy does appear to propagate in a smoothly continuous fashion. It is only when we "look more closely," examining the microscopic properties of light, that we notice its discontinuous, quantized grain.

It took a generation for the truly revolutionary implications of quantum mechanics (QM) to become clear. Under the lingering influence of classical thinking, it was natural to assume that the discontinuity of energy was not really fundamental. For, if the properties of a quantum of energy were to be subject to complete scientific determination, it seemed as if the discontinuity ultimately had to be reducible to continuous expression via an underlying space-time substrate. Yet, by 1930, most physicists had arrived at the conclusion that no such reduction is possible. At this point, the majority of researchers felt obliged to accept the idea that Planck's microscopic quantization implies a basic indivisibility of energy that confounds analytic continuity and, in so doing, calls into question all classical thinking about space and time, including that of Einstein. Therefore, in light of quantum mechanics, "the concepts of spatial and temporal continuity are hardly adequate tools for dealing with the microphysical reality" (Čapek, 1961, p. 238).

The microscopic loss of continuity may be better understood by considering more closely Planck's constant. This number in fact may be expressed as an *action*. The value h is transformed from a mere numerical quantity into "spin," that is, internal angular action (momentum), by the application of *phase*, as given in the formula h/2Π = ħ. Here h is operated upon by a phase of 2Π radians, equivalent to a turn of 360°. In quantum mechanics, ħ is termed the *quantum of action*. It is an indivisible "atom of process," one not reducible to smaller units that could be applied in its quantitative analysis. Thus, at the sub-microscopic Planck threshold of 10^{-35} meter, the analytical continuity of space gives way to a "graininess" or discreteness that admits of no further quantitative determination. We see here the intimate relationship between the indivisibility of the quantum domain and its basic indeterminacy or uncertainty. According to the well-known uncertainty principle postulated by Werner Heisenberg in 1927, there is a built-in limit to the information we can obtain about the physical properties of quantum systems. This limitation can be stated in terms of Planck's constant: $\Delta p \Delta q \approx \hbar$, where p and q are variables such as position and momentum, or time and energy (variables that are paired or conjugated so as to be essentially indivisible from *each other*). The formula says that the product of the uncertainties (Δs) of such paired terms approximately equals (cannot be less than) the value of Planck's constant. Clearly then, the phasic indivisibility (h/2Π) of Planck-level action is equivalent to its uncertainty (ΔpΔq).

There is another way to look at the quantum uncertainty. Nearing the sub-microscopic Planck length, it appears that precise objective measurement is thwarted by the fact that the energy that must be transferred to a system in order to observe it *disturbs* that system significantly. This well-known "problem of measurement" in quantum mechanics expresses quantum indivisibility in terms of the indivisibility of the *observer and the observed*. It seems that, in QM, the observer no longer can maintain the classical posture of detached objectivity; unavoidably, s/he will be an active participant. Evidently this means that quantum mechanical action cannot be regarded merely as objective but must be seen as entailing the intimate *blending of subject and object*. Of course, the blending of subject and object is just what had been hinted at in Michelson and Morley's experiment with light—a blending that Einstein denied in order to maintain objectivism. We will see shortly how the practitioners of quantum mechanics reacted to the threat to objectivism that confronted them.

The ultra-microscopic research on light appears to hold the same implications as did the research of Michelson and Morley: Moving down the scale of magnitude to the Planckian threshold, the classical container becomes "leaky" and *apeiron* resurfaces. The classical expectation, rooted in the assumption of continuity, is that scale shrinkage eventually brings us to the null volume of the dimensionless point. At this level of nature, all material bodies lose their extension and vanish. As a consequence, space becomes vacuous and cold, devoid of energy or matter. Quantum mechanics presents a dramatically different picture of microphysical reality. Instead of energetic nature being totally subdued, she returns with a vengeance. As the philosopher-physicist David Bohm put it, in classical physics, "the field changes over very short distances are negligibly small," whereas, "in the quantum theory…the shorter the distances one considers, the more violent are the quantum fluctuations associated with the 'zero-point energy' of the vacuum. Indeed, these fluctuations are so large that the assumption that the field operators are continuous functions of position (and time) is not valid in a strict sense" (1980, p. 85).

The vacuum fluctuations that disrupt classical continuity are of course associated with Planck's constant, which embodies the indivisibility of action, irreducible uncertainty, the inseparability of subject and object. This wild variability of nature in the small is clearly reminiscent of the "irrational motion," the "fleeting potencies and constantly changing tensions" (Graves, 1971, p. 71) of Plato's tenuously continuous receptacle. We have seen that, with the Renaissance, Platonic proto-space was superseded by a new order of space, one whose continuity was enhanced. Just as this order was subsequently thrown into doubt by the Michelson-Morley experiment on the velocity of light, so too was it called into question by Planck's blackbody research. Can we say that, whereas Einstein attempted to plug the hole in the spatial vessel by denying it (via his proposal of a four-dimensional space-time continuum), Planck and his successors *accepted* the discontinuity? Did quantum physics give up Einstein's effort to uphold objectivism? Did it embrace the indivisibility of object and subject?

We have explored what was at stake with the dawning of the Age of Light. The mid-nineteenth century threat of discontinuity and concomitant loss of objectivity constituted nothing less than a threat to the life of the *cogito* itself. To accept *apeiron* would be to accept the collapse of the projection upon which human identity had come to depend. It is therefore not surprising that Planck could accept *apeiron* no better than could Einstein. The fact is that quantum mechanics did not simply give up on the continuum. Instead, implicitly, the attempt was made to retain continuity through an approach that is even more abstract than Einstein's.

Let us consider a central feature of the quantum theoretic formalism: *analysis by probability*. According to the classical ideal, the extensive continuum is infinitely differentiable, which means that the position of a system within it is always uniquely determinable. When QM was confronted with the *inability* to precisely determine the position of a particle in microspace, it certainly did not just resign itself to the lack of continuity that creates this uncertainty. Instead of allowing the conclusion that a microsystem in principle cannot occupy a completely distinct position—which would be tantamount to admitting that microspace is not completely continuous, a multiplicity of continuous spaces was axiomatically invoked to account for the "probable" positions of the particle—"it" is locally "here" with a certain probability, or "there" with another. This collection of spaces is known as *Hilbert space*. *N*-dimensional Hilbert space plays a role similar to that played by Einstein's four-dimensional space-time continuum: it responds to the threat of discontinuity by restoring continuity through an act of abstraction. And, as with Einsteinian relativity, the quantum mechanical abstraction of classical space brings with it an abstraction of subjectivity.

We have discovered that relativity theory implicitly transforms the old subject into an object cast before the analytic gaze of a more abstract, higher order subject. The same can be said for quantum mechanics. In effect, the quantum mechanical analyst assumes a super-ordinate vantage point from which he or she is able to consider alternative acts of classical observation and weight them probabilistically, with each act corresponding to a different subspace of the Hilbert space. That the QM analyst is not so much observing objectively existing particles but the *observations* of particles, is implicit in the attitude of the founding father of quantum mechanics. Max Planck's departure from classical objectivism is especially evident in the following remarks:

> "[W]e have no grounds on which to make any sort of comparison between the actualities of external effects and those of the perceptions provoked by them. All ideas we form of the outer world are ultimately only reflections of our own perceptions. Can we logically set up against our self-consciousness a 'Nature' independent of it? Are not all so-called natural laws really nothing more or less than expedient rules with which we associate the run of our perceptions as exactly and conveniently as possible?" (Planck quoted in Jahn and Dunne, 1984, p. 9)

If it seems that Planck was merely advocating subjectivism in expressing these sentiments, that was certainly not the case. At bottom, Planck was not saying that "all is subjective," any more than Einstein was saying that "all is relative." By no means was Planck urging the acceptance of *apeiron*. The idea was not that, in surpassing the old objectivism, the quantum mechanical subject becomes *fused* with the objects s/he observes. On the contrary, detached objective analysis was still the order of the day. Only now, the "objects" to be analyzed were not concrete substances but the "run of our perceptions." If the quantum mechanical analysis of perception was not to dissolve into a "viciously circular subjectivism," if "scientific objectivity" was to be maintained, the perceptions of the *analyst* of perception themselves had to be *exempted* from the analysis. That is to say, two distinct, categorially separate levels of perceptual activity had to exist: that which was to be analyzed, and that through which the analysis was to take place. The former was constituted by the old observational activity that was now objectified within the framework of the Hilbert space, whereas the latter corresponded to the more abstract, higher-order, wholly implicit perceptual activity of the quantum mechanical "observer" standing outside of Hilbert space. It is clear that this quantum mechanical subject assumed the same detached, "purely objective" stance as did his Cartesian predecessor. Still operative in its essential relations was the basic formula of object-in-space-before-subject.

Just how effectively does Hilbert space preserve continuity and the abstract objectivism with which it is coupled? Each subspace of the multi-space expression is made continuous within itself to uphold the mutual exclusiveness of the alternative positions of the observed particle. Such subspaces must be disjoint with respect to each other, their unity being imposed externally, by fiat, rather than being of an internal, intuitively compelling order. So, in the name of maintaining mathematical continuity, a rather extravagantly *dis*continuous state of affairs is actually permitted in the standard formalism for quantum mechanics, an indefinitely large aggregate of essentially discrete, disunited spaces. This approach is reminiscent of an earlier stratagem for resisting change.

Bohm (1980) describes the universally sanctioned response given at a time in history when the order of thinking and perceiving that pre-dated the Renaissance became suspect. The order of ancient Greece was predicated on the idea of the circle, expressed specifically in the conviction that celestial bodies trace perfectly circular orbits. "To be sure," noted Bohm,

> when more detailed observations were made on the planets, it was found that their orbits are not actually perfect circles, but this fact was accommodated within the prevailing notions of order by considering the orbits of planets as a super-position of *epicycles*, i.e., circles within circles. Thus one sees an example of the remarkable capacity for adaptation within a given notion of order, adaptation that enables one to go on perceiving and talking in terms of essentially fixed notions of this kind in spite of factual evidence that might at first sight seem to necessitate a thorough-going change in such notions. (1980, p. 112)

I might add that, in the Ptolemaic notion of "epicycles," there was a kind of mockery of the image of the circle, an extravagantly complex, wholly gratuitous replication of this image, one that did not truly achieve what it sought, given the fact that the circle actually had lost its original meaning and effectuality. Yet those who were determined to uphold circularity blinded themselves to this.

Do not the self-continuous subspaces of the Hilbert formalism mock continuity in largely the same way the self-circular epicycles of antiquity mocked circularity? The mid-nineteenth century challenge to the status quo was formidable indeed. What was being called into question in the Age of Light was not merely an objectified image that the subject can reflect upon, but the reflective posture itself, the whole relation of object-in-space-before-subject. And the response to this profound questioning has been decidedly Ptolemaic. In the method that has been adopted, rather than offering a novel approach that would be genuinely amenable to the discontinuous, non-objective phenomena of the microworld, the attempt is made to pour the "new wine" into an "old bottle." The standard stratagem is to express the upsurgent discontinuity in such a way that continuity is implicitly preserved. Yet it is evident that continuity actually is not successfully maintained in quantum mechanics; despite the artificial semblance of it, on the subtler level of the form that quantum theory takes, continuity is denied (Rosen, 1983, 1994). (Also relevant to this issue are Bohm's [1980] and Josephson's [1987] critiques of mathematical formalism, and the critiques of Stapp [1979] and Gendlin and Lemke [1983].) As a matter of fact, if we follow the full development of quantum mechanics, we see that, in the end, even the semblance of continuity is lost.

Hilbert space does not retain its quasi-continuity for all levels of energy at all scales of magnification. Its range of applicability is limited to the comparatively low-energy regime that lies above the Planck length of 10^{-35} meter. It is true that, in studying the phenomenon of electromagnetic radiation, Planck brought us into an energy domain in which classical continuity was shaken. But while the domain in question is surely microscopic and Planckian uncertainty becomes a significant factor here (whereas, in the large-scale classical world, it does not), this realm of interaction remains considerably above the *ultra*-microscopic, ultra-energetic Planck scale.

However, in the course of the twentieth century, physicists probed the microworld ever more deeply. This permitted them to advance their project of arriving at a unified understanding of nature. Whereas the fundamental forces of nature appear very different from one another at lower energy levels (and orders of magnification), physicists, by pushing their quantum mechanical research into the high-energy sub-microscopic domain, could now account for intra-nuclear forces (the so-called strong interactions), along with subatomic decay forces (weak interactions), in a manner that successfully unified them with each other, and with the electromagnetic force. Yet, in drawing closer to the Planck length of 10^{-35} meter, the element of uncertainty grew greater.

To complete its quest for unity, physics now faces one final task. It must include in its quantum mechanical analysis the one force of nature hitherto unaccounted for, namely, gravity. The problem is that gravitation, unlike the non-gravitational forces, resists QM treatment until the bitter end. That is, gravitational energy behaves classically, appears to retain its continuity all the way down the scale of magnitude to the Planck length itself. It is precisely here that a QM theory of gravitation would have to operate to fulfill its aim of total unification. Of course, the Planck length is the threshold at which spatiotemporal turbulence goes out of control and uncertainty becomes all consuming. Crossing this threshold, the quasi-continuity of Hilbert space yields to utter discontinuity. Not even a probabilistic analysis of nature is possible here, as is reflected in the infinite probability values obtained for equations dealing with sub-Planckian reality.

What we are now better able to see is the intimate relationship between quantum mechanics and Einstein's theory of gravitation. It is not merely that the former reaches its Planckian limit and encounters irrepressible discontinuity in a manner that is *analogous* to the black-hole limitation confronting the latter. For what may seem at first like analogous but different limitations, actually can be said to constitute the very *same* limitation. The work of physicist Arthur Eddington (1946) contributed to an understanding of this. In his own effort at unification, he demonstrated the equivalence of relativistic curvature with quantum mechanical uncertainty:

> Curvature and [quantum mechanical] wave functions are alternative ways of representing distributions of energy and momentum....We have introduced the curved space of molar relativity theory as a mode of representation of the extraordinary fluctuation, and have obtained the fundamental relation (3 · 8) between the microscopic constant σ and the cosmological constants R_0, N. (1946, p. 46)

Bearing in mind the essential equivalence of uncertainty with *discontinuity*, Eddington's findings are consistent with the fact mentioned earlier that, the greater the curvature of space-time, the closer we approach to the loss of continuity realized in the singularity of the black hole. Therefore, the production of curvature in general relativity, which culminates in the infinite warping of space-time found in the heart of a black hole, maps onto the production of Planckian uncertainty, the significance of which progressively increases as we descend into the microcosmos. It seems then that the black hole singularity of general relativity is none other than the manifestation of quantized gravitational energy at the Planck length.

It is true, however, that general relativity and quantum mechanics do not converge explicitly until the "bitter end." Before the curvature of space-time actually reaches an infinite value in the black-hole singularity, gravitational uncertainty is only implicit, is not expressed overtly. This limitation can be clarified by considering the question of scale. Are not black holes *large-scale* phenomena, astrophysical events taking place at the opposite end of the scale of magnitude from the microphysical happenings of quantum mechanics? On the contrary, upon entering the singularity of the black hole, the pervasive uncertainty about distance that arises here (owing to the loss of continuity) renders any notion of "large-scale" vs. "small-scale" inoperative. Simply stated, the scale of magnitude collapses. However, so long as the curvature remains finite, its increasing value does not show itself as an out-and-out collapse of scale, but only as a *reduction* of scale accompanying the collapse of *matter*. As a massive stellar body grows older and runs out of hydrogen fuel, the balance of energy is tipped toward gravitation and the attraction of the star's constituent particles for each other gains supremacy, causing the star to implode, to cave in upon itself. This high-density compression of the star leads to a drastic reduction in its volume. The effect is enhanced as we move toward the center of the black hole. Compression becomes so great that we eventually reach the ultra-microscopic Planck threshold! This is the point below which space-time curvature does become infinite and the singularity appears. It is not until we cross this threshold that the microscopically scaled phenomenon becomes one that is utterly unscaleable. Only now is the gravitational uncertainty explicitly manifested, making its presence felt in such a way that it cannot be controlled.

Einstein had hoped to steer clear of microphysical uncertainty. "God does not play dice with the universe," were his famous words. But his hope proved to be vain. The crisis besetting Einstein's theory is that its end result implies a radically discontinuous "microstructure" that blatantly contradicts its initial premise of an entirely continuous one. Einstein's "macroscopic" theory comes to an end at the very same place where quantum mechanics ends: at the Planckian limit at which the quantum of action gains full sway. The singularity at the center of the black hole—far from being a "macroscopic" phenomenon that is merely *analogous* to "microscopic" uncertainty —is that self-same uncertainty. In the final analysis, we cannot distinguish between "large-scale" and "small-scale" forms of uncertainty simply because uncertainty entails the collapse of scale.

Thus, in the singularity of the black hole, relativity theory and quantum mechanics come together. Of course, this "unification of the field" is hardly what science had intended, since the unity is realized in negation, marking as it does the failure of determinative analysis. It is true that, as quantum mechanical research advanced through the twentieth century, a measure of tolerance for uncertainty was built up among physicists. This is the reason Einstein's theory was not simply abandoned with the late twentieth century confirmation that black holes do exist. Had Einstein himself survived to witness the confirmation, it is unlikely he would have been so blasé, for uncertainty was completely anathema to him, even in the smaller doses with which quantum mechanics had inoculated itself. We have seen that Einstein's theory, like quantum mechanics, was essentially an attempt to mitigate the threat of discontinuity brought on by the phenomenon of light. Einstein's way of doing this was to accommodate electromagnetism by extending the classical continuum. His attitude was that continuity must be preserved in a relatively concrete way, without too great a compromise to classical intuition. Yes, the old notion of three-dimensional space plus time now did have to be replaced by the four-dimensional space-time continuum. Nevertheless, the intuition of a unitary geometric continuum was to be kept intact. Quantum mechanics, for its part, did not seek to dispel discontinuity in this concretely intuitive fashion but gave itself the license to proceed *counter*-intuitively; it *allowed* microphysical discontinuity, though formalistically imposing upon it an abstract continuity, as noted above. Therefore, whereas Einstein was completely repelled by discontinuity, quantum mechanics came to take it with greater equanimity.

Yet there is clearly a limit to QM's tolerance of discontinuity. The line is drawn at the Planck length, the threshold below which discontinuity becomes totally unconstrainable. And this is the line at which physics stands today. To complete its "theory of everything," its unified account of the diverse forces of nature, it must now come to terms with gravitation, and this brings it to the Planck length—to the brink of utter chaos, that is. Yet by no means is science prepared to allow its long drive toward unification to be thwarted just when "victory" appears to be in sight. Emboldened by its past success at managing uncertainty, intoxicated by the heady prospect of taming Mother Nature once and for all, contemporary physics seeks ever new techniques for achieving its rational unification, one in which continuity is ultimately retained. A prime example of this is found in an approach to unification known as *string theory*. Since this theory has excited widespread interest in the physics community and has resulted in a prodigious amount of research, I will consider it in some detail.

The basic idea is: *What we can't know can't hurt us*. That is, if it is impossible for us to *probe* the cataclysmic energies allegedly prevailing below the Planck length, then we are entitled to assume that said energies either actually do not exist, or that they exist but cannot influence us, and, therefore, can be completely disregarded. The physicist Brian Greene—in his attempt to explain "how string theory calms the violent quantum jitters" (1999, p. 158), "tames the sub-Planck-length quantum undulations of space" (p. 157)—acknowledges the positivistic nature of this strategy: "A positivist would say that something exists only if it can—at least in principle—be probed and measured" (p. 156). Because "the violent sub-Planck-length…

fluctuations cannot be measured…according to string theory, [they] do not actually arise" (pp. 156–157).

But how can we say that there is no sub-Planckian world, or that this world does not affect us? Do not the infinities that arise in equations seeking to unify quantum mechanics and general relativity attest to the influence of the ultra-microscopic energies? String theory would answer in the negative. It would say that the troublesome infinities simply come from the pre-string theoretic *assumption* of ultra-microscopic energies. The equations that had been written to successfully unify the three non-gravitational forces had implied that elementary particles are point-like; having no *minimum* length, in principle they could be probed below the Planck length. It is when this assumption is maintained in the attempt to include gravity that we are brought to the sub-Planckian scale where the equations go infinite. To remove the obstacle to formulating a successful theory of quantum gravity, string theory offers an alternative assumption about the basic constituents of the universe. Instead of being unextended, point-like particles, they are *strings*: "tiny, one-dimensional filaments somewhat like infinitely thin rubber bands, vibrating to and fro" (Greene, 1999, p. 136). How long is a string? It is roughly the size of the Planck length. By supposing the string to be the *basic* constituent of nature, string theory precludes any movement *below* the Planck length, into the chaotic realm where infinities would arise.

Now, while Greene is a strong proponent of string theory, he admits that the "cosmological positivism" inherent in this approach (—what we can't know can't exist and therefore can't hurt us—) may seem like nothing more than a trick, a conceptual "sleight of hand….Instead of showing that string theory tames the sub-Planck-length quantum undulations of space, we seem to have used the string's nonzero size to skirt the whole issue completely" (p. 157). Having thus raised the question of whether string theory has any real substance, Greene offers a couple of comments intended to suggest that it actually does. However, in his first remark, he only seems to compound the problem.

Here Greene repeats his point that it is not string theory that lacks substance but the infinities of the pre-string theoretic standard model, which are seen to be "an artifact of formulating general relativity and quantum mechanics in a point-particle framework" (p. 157). But is this critique of the standard model followed by a positive demonstration that string theory, for its part, does substantively correspond to reality? On the contrary, Greene merely informs us that string theory furnishes "new rules" that help us play the "game" of eliminating infinities more effectively than did the rules of the earlier model: "String theory tells us that we encountered these problems [of infinities turning up in our equations] only because we did not understand the true rules of the game; the new rules tell us that there is a limit to how finely we can probe the universe" (p. 157). Then is it just a game after all, one whose rules can conveniently be adjusted to obtain the desired results?

The second comment offered by Greene to demonstrate that string theory cannot be dismissed as mere sleight of hand is that it has proven to be

> consistent with the most basic physical principles such as conservation of quantum-mechanical probability (so that physical objects do not suddenly vanish from the universe, without a trace) and the impossibility of faster-than-light-speed transmission of information....The truly impressive feature of string theory is that more than twenty years of exacting research has shown that although certain features are unfamiliar, string theory *does* respect all of the requisite properties inherent in any sensible physical theory. (1999, p. 158)

Nevertheless, while the theory generally conforms to what would be expected of "any sensible physical theory," this is a far cry from providing a specific set of infinity-free equations for quantum gravity. Thus far, it is only *in principle* that string theory eliminates the infinities found in the standard model. Accordingly, Greene is quick to acknowledge that, since the theory has not yet reached a mature stage of development, the jury is still out on it, despite the great promise he believes it holds for unifying the forces of nature. Moreover, even if string theory *could* arrive at a clear-cut theoretical description of quantum gravitational interaction, it would be impossible to test the theory in a concrete way, since the test would have to be conducted on a scale "some hundred million billion times smaller than anything we can directly probe experimentally [!]" (Greene, 1999, p. 212).

Now, Greene emphasizes that "the spatially extended nature of a string is the crucial new element allowing for a single harmonious framework incorporating both [gravitational and quantum mechanical] theories" (1999, p. 136). If string theory had proven itself in a substantive and convincing way, it might seem that there would be little point in questioning this characterization of the string. However, given the incompleteness of the theory and the grave difficulties entailed in its external validation (via experimental observation), it appears reasonable to inquire into the *internal* validity of the idea. Does the notion of a fundamental particle with finite extension really make any sense?

Greene poses the question, "What are strings made of?", and answers it as follows:

> [S]trings are truly fundamental—they are "atoms," *uncuttable constituents*, in the truest sense of the ancient Greeks. As the absolute smallest constituents of anything and everything, they represent the end of the

> line....From this perspective, *even though strings have spatial extent* [my emphasis], the question of their composition is without any content. (1999, p. 141)

Is there not a contradiction here? Did we not find earlier, in examining the classical concept of spatial continuity (Section 1), that, to be spatially extended is to be *cuttable*, in fact, infinitely divisible? How then could a string be a fundamental particle, an indivisible ingredient of nature, when it is spatially extended? Perhaps, when Greene speaks of the string as he does, he has some unusual definition of "spatial extension" in mind. If that is the case, we are given no hint of it. Consequently, we are left with the default meaning of "extension," the classical intuition of it that entails the idea of being *divisible*. The string, then, is an indivisible particle that is divisible—a contradiction to be sure. Let us look more closely at why string theory is forced into contradiction.

Within the classical framework of object-in-space-before-subject, the idea of an indivisible, point-like atom or particle is indispensable. In order for the subject to analyze effectively an object in space, s/he must be able to determine what the object is "made of," what its basic constituents are. The common tendency is to reify the atom; we think of it as a "tiny object" contained within the larger object that is constituted by it. But atoms are not "contained" in the same manner as ordinary objects are (matches inside a box, for example). For there is actually a *categorial difference* between the atomic constituent of an object and the object itself. That is because, by definition, the "atoms" of which the object is composed, being indivisible, are not themselves open to analysis. The idea is not to say what *they* are made of; rather, they are the *means* by which the analysis of the larger object is performed. Therefore, the point-like elementary particles, instead of belonging with the object term of the classical formula, in fact belong with the subject term. The fundamental particle to which an object is reduced is the element upon which the subject's analysis of that object is built.

We can say that the particle is the locus or point-origin of the *cogito's* conceptual analysis—much as the *cogito's perceptual* activity is centered by the 0,0 origin of Cartesian space or Minkowskian space-time, as discussed in Section 2A. In our previous examination, the 0,0 center of the continuum was characterized as the locus at which observation begins, as surrogate for the subject's "eyes," as the point from which her perspective upon the objective world opens out. Does the idea of the point *particle* enter this perceptual account, as it does in conceptual analysis? Contemporary physics tells us that we observe ordinary objects by bouncing photons off them, which then reach the retinas of our eyes. In this perceptual activity, the photons serve as elementary particles. It is not they that are perceived; rather, these microscopic "atoms" are the means by which the observer perceives the object. The photon thus plays the role of surrogate for the subject. As the elementary probe through which observation takes place, the photon is the point center of the Einstein-Minkowski coordinate system. In sum, whether we are speaking of the perceptual or conceptual operations that the subject performs upon its objects, the point particle is of central importance.

As Greene notes, the standard model of unification maintains the classical assumption of point particles. Of course, the irreducible uncertainty inherent in Planck's constant exposes a basic limitation of this assumption that previously had not been evident. To restate the Planckian restriction a little differently from the way I put it above, it indicates that, the closer we come to the point-like atomic origin of observation and analysis, the fuzzier that origin becomes. Yet, in *non-gravitational* quantum mechanics, the fuzzing out of the 0,0 origin is tolerable. We have seen that contemporary physicists have been content to speak of microscopic measurements in *probabilistic* terms, thereby acknowledging that observation and analysis cannot be fully determinative activities. The prevailing philosophical attitude toward this constraint is given in the positivistic pragmatism of the Copenhagen Interpretation of quantum mechanics. Although there is indeed a built-in limitation to specifying the physical states of a microsystem (say, the position and momentum of an individual particle), the probabilistic analysis can be used with unprecedented exactitude to determine the behavior of macrosystems (whole ensembles of particles). Since the microsystems in question are so miniscule, their uncertainty can be overlooked in the macroscopic operations of the laboratory. Therefore, using Planck's constant, the work of science can proceed quite effectively. But do we not have to acknowledge a complete loss of precision at the ultra-microscopic level of nature? Should this not be of concern to us? Not necessarily, says the Copenhagen Interpretation. By taking microscopic imprecision into account, a high degree of macroscopic precision can be gained. Since the calculations work, there is actually no need to trouble ourselves with the meaning of the underlying uncertainty. What we don't know isn't hurting us in the practical activity of science that we engage in; therefore, we don't have to think about it.

So it is not only string theory that assumes a positivistic stance. The standard model of unification and the Copenhagen Interpretation of quantum mechanics on which it is based also adopt this posture. However, whereas the latter two can afford their positivism, it seems that string theory actually cannot, since *its* positivism must be bought at the price of self-contradiction.

Positivism becomes exorbitantly expensive when gravitation must be included in the unified account of nature. Given the colossal ultra-microscopic energies required for this, even Ptolemaic Hilbert space collapses, since finite probability values no longer can be assigned to terms that describe the physical states of particles. Below the Planck length then, twilight becomes utter darkness. So it seems that, in the case of quantum gravity, what we don't know *does* hurt us. String theory's attempt to circumvent this problem and maintain positivism as a viable approach leads to the apparent contradiction we have discussed: an elementary particle with finite extension—that is, a point particle that is *not* a point particle. The particle *cannot* be a point if the ultra-microscopic infinities are to be avoided. And yet, the particle *must* be point-like, it must be indivisible, if an objective analysis is to be performed. For, if the particle is not elementary, if it does not have an unextended atomic character, analytical activity cannot be centered by a 0,0 origin in any definitive fashion.

There is a sense in which the contradiction of string theory throws us right back to the dilemma that was implicitly faced by Michelson and Morley a hundred years before: In attempting to treat light as an object, a *fusion* of object and subject was disclosed that foreshadowed the collapse of classical identity. Until the last quarter of the twentieth century, physics naturally had done its utmost to suppress this implicit fusion. The standard model of unification had still been able to assume that the fundamental particle is simply unextended; therefore, the separation of subject and object could continue to be upheld, at least in probabilistic approximation: the point particle is the 0,0 origin of the subject, and the extended entities that are observed and analyzed by the subject are its objects. But then, around 1975, the need to deal with quantum gravity became a pressing issue. Enter here string theory, with its "extended atom." Unlike the point particle of standard unification theory, the string evidently must be *both* indivisible *and* spatially extended—in effect, *both subject and object.* It is this implication that subverts the posture of object-in-space-before-subject that physics had long sought to maintain in its quest for unification.

What the contradiction at the heart of string theory discloses is that, in the final analysis, *apeiron*—"the inchoate flux of…contraries"—cannot be denied. Does this preordain a simple regression into chaos? I suggest that it does not. I propose that the apparent self-contradiction of string theory in fact might entail a certain dialectic of transformation.

It is clear by now that *apeiron* had never been truly eliminated in the *cogito's* strivings for individuation. Instead, the chaotic element in nature merely had been relegated to the background via the idealizations of science. In the contradiction of string theory, *apeiron* now enters the foreground in an irrepressible way. Yet I submit that the attendant threat of wholesale regression lies not so much in *apeiron* per se, but in the continuing attempt to deny her. It is in resisting what actually can be resisted no longer that catastrophe threatens. The ongoing resistance is certainly understandable. To the *cogito* who clings to his detached stance, *apeiron* can only signify the complete loss of individuality. But it is surely not the case that the *cogito* had ever *fully achieved* individuality (his idealized projections to the contrary notwithstanding). What we will see in Part Two of this book is that it is precisely when *apeiron* is *accepted* that the process of individuation can be completed in earnest. By no means does such an acceptance involve passively resigning oneself to dissolution in the chaotic multiplicity of *apeiron*. We shall discover that one can actively participate in the many of *apeiron* without *losing* oneness, and that, in fact, oneness (unity, individuality) can genuinely be achieved *only* through said participation. What is called for here is a *dialectic* of the one and many that favors neither the unity of fixed form nor formless multiplicity. A dialectic of *trans*formation is required.

Were this approach to be taken to the blending of subject and object found in string theory, an unacceptable contradiction could become a fruitful paradox. Elsewhere (Rosen, 1988, 1994), I attempted just such a dialectical reformulation of the dimensional underpinnings of the theory. Carrying this out effectively involves nothing less than a radical transformation of the foundations of physics. Specifically, the Baconian enterprise of subduing Mother Nature, imposing unity upon Her from afar, is supplanted by an endeavor in which we seek to understand Nature from within, by participating freely in Her flowing diversity (see Fox-Keller, 1985). In the dialectical rendering of "unified field theory," while unity does not simply dissolve, neither does it prevail in the one-sided terms of the *cogito*. Instead there is a *unity-in-diversity*, a *diversity-in-unity*. Indeed, this is the mark of individuation fulfilled, as we are going to see. In my original plans for the present book, string theory was to be revisited in a chapter that would carry forward my past work on the subject, and relate it to the concerns of this volume. In the end, however, I dropped the idea, since it became clear to me that the task of doing proper justice to a dialectical unified field theory in a book not primarily devoted to the topic would be too unwieldy. So, further exploration of this intriguing and weighty matter must be left for the future. Nevertheless, the account that will unfold in the chapters to follow—in systematically elucidating the dialectic of subject and object, of space, time, and dimension—will not only shed light on questions of crucial concern to physicists and philosophers alike, but will provide the needed foundation for future efforts at unifying the field.

Two

THE CRISIS OF DISCONTINUITY IN THE BROADER CULTURE

It is clear from the writings of contemporary philosophers as diverse as Martin Heidegger (1962/1977), Ludwig Wittgenstein (1953), Tanabe Hajime (1986), and Michel Foucault (1975), that science is first and foremost a human enterprise. As such, it reflects the changing cultural context within which it is enmeshed. It was hardly just physics that confronted a crisis of discontinuity in the middle of the nineteenth century. Upheavals of this kind were evident in almost every sector of human life. In this chapter, I am going to consider two major cases in point. In preparation, let me return briefly to the comparison of contemporary theoretical physics with the classical physics that preceded it.

Recall Gebser's characterization of the new world of the Renaissance as the "*perspectival* world" (see p. 11). We have found that the move to the Renaissance entailed a step in development toward a more focused view of reality, one in which the world was now experienced from the detached perspective of a more fully individuated subjectivity. From this viewpoint, objects took on greater dimensionality; they appeared in *depth*—as is reflected in the three-dimensional quality of Renaissance art not evident in the flat-looking works of the Middle Ages. We also know that, in classical physics, the principle of Galilean relativity rendered all perspectives interchangeable. Different points of view could readily be transformed into one another because these local perspectives were implicitly interconnected by being grounded in the newly consolidated spatial continuum over which the *global* perspective of the *cogito* was assumed to preside. Of course, in post-Renaissance physics, both the new order of subjectivity and the new order of space that accompanied it remained implicit. It was not they to which the attention of science was directed. Subjectivity itself was not observed, but was the changeless transcendent viewpoint from which *objects* were observed, measured, and transformed. Space per se was not observed, but was the immutable medium through which the *cogito's* operations upon material objects (matter and energy) were carried out.

Then came the nineteenth century upsurgence of *apeiron*. We have seen that this constituted a fundamental challenge to the classical formulation of object-in-space-before-subject. We know as well what science did to meet this challenge. The discontinuity of space associated with the study of light seemed to suggest that classical space no longer could be taken *a priori* as a changeless medium. Einstein responded to this by *empiricizing* three-dimensional space, treating it, in effect, as a variable *object* in a new, more abstract, *four*-dimensional medium. In quantum mechanics, even greater heights of abstraction were reached in responding to the challenge of discontinuity posed by the phenomenon of light. The *n*-dimensional array of subspaces that constitute the Hilbert space is certainly a far cry from the singular Cartesian continuum.

Closely related to the discontinuity intimated by the investigation of light was the fact that what was being observed could no longer cleanly be separated from the act of observation itself, which meant that the subject could no longer maintain its detached anonymity. The response of modern physics was to treat the Cartesian subject as an object (an "observational event") cast before a new, even more abstract, still anonymous subject. Thus, in order to maintain the formula of object-in-space-before-subject, what had previously constituted space and subjectivity were now being treated as *objects* in a more abstract space, before a more abstract order of subjectivity. When these novel objects were fashioned, what became of the *old* sense of the object? When attention was withdrawn from the relatively concrete objects of the Mechanical Age and re-invested in the more abstract spatial and observational "objects" emergent in the Age of Light, what happened to the Cartesian sense of perceptual depth and perspective? The answer is that it collapsed. Pre-Einsteinian differences—differences in perspective that did not explicitly take into account the observer and his space as higher order objects—no longer made a difference. To the eye of the modern physicist, the old world of depth thus looked distinctly flat. In sum, we may say that, in attempting to meet the challenge of *apeiron* in the Age of Light, a more abstract mode of experience arose in science, one in which attention shifted from the concrete objects perceived to the act of perception and its space as new kinds of objects; in the process, there was a flattening of Cartesian depth as the earlier sense of perspective was relinquished.

Just such a transformation of experience was also witnessed in the field of *art*. Art historians Paul Vitz and Arnold Glimcher (1984) identify these very same features—abstraction, flattening of perspective, the shifting of attention to the act of perception itself—as primary characteristics of modernist art.

According to Vitz and Glimcher, abstractness is expressed in modernist art by the reduction of recognizable content to formal geometric elements (Mondrian's fields of vertical and horizontal lines, for example), and, in the extreme, by the complete elimination of content, leaving only the blank painting surface itself (as in the ultra-minimalism of Malevich). Regarding the second feature of modernism, its flatness, the authors observe that the effect of a flattened, non-perspectival picture plane arises from the artist's recognition of "the painting's surface and in turn the canvas as a real object rather than an illusionistic 'window'" (1984, p. 5). Of course, treating the canvas itself as the reality is an act of abstraction, so the flattening of the picture may indeed be said to result from this tendency to abstraction. And both of these characteristics may be related to the third essential trait of modernism: its "perceptual character" (Vitz and Glimcher, 1984, p. 7).

Beginning with Impressionism, art became a subjective, self-conscious endeavor. In the work of Claude Monet, Pierre Renoir, Camille Pissarro, and others, the interest of artists no longer was wholly invested in the objects they perceived, but was concerned with their own impressions of those objects, their own perceptual processes. This withdrawal of interest from the classical object contributed to the flattened appearance of even the earliest Impressionist works (see, for example, Vitz and Glimcher's comparison of Manet's *Olympia* with Titian's *Venus of Urbino* [1984, pp. 48–49]). But modernism did not limit itself merely to exploring perceptual process; in advancing from Impressionism to Cubism, it went on to *objectify* perception. We might say that, whereas Impressionism raised the *question* of subjectivity, with Pablo Picasso and Georges Braque that question was "resolved" by transforming the subject into an object.

I propose that the Impressionist experiment with subjectivity is related to Cubism in a way that parallels the relationship of the Michelson-Morley experiment to Einsteinian physics. It was in the same decade in which the work of Michelson and Morley posed its implicit challenge to Cartesian subjectivity and classical space that a corresponding challenge was being issued in the field of art. I will take a painting by Edouard Manet as my case in point: *Bar at the Folies-Bergère*, exhibited in Paris in 1882 (see next page).

Vitz and Glimcher state that Manet's *Bar* is an example of *fractured space*: "two or more discrete lines of view [are] present at the same time in a given portrayal of space: these separate but simultaneous views break or fracture what was once (seen as) homogenous" (1984, p. 118). In *Bar*, the left two thirds of the picture give a view of the woman from a frontal perspective, whereas the mirror image of the woman's back and the top-hatted gentleman shown in the right third of the painting suggest an angle of vision that is displaced to the left of center. Vitz and Glimcher observe that this juxtaposition of opposing perspectives leaves the viewer in an ambiguous state. The undercurrent of visual tension that accompanies the viewing of *Bar* arises from the fact that when attention is occupied by one perspective, the pull of its overlapping opposite is subliminally felt. A significant effect of this perceptual uncertainty is to disrupt the viewer's involvement with the object perceived and shift his or her attention to the act of perception itself.

Figure 3. Edouard Manet's *Bar at the Folies-Bergère* (1882)
Courtauld Institute Galleries, London.

As with the Michelson-Morley experiment that was to take place five years later, Manet's fracturing of classical space brought subjectivity out of the background and into the open, creating the ambiguous blending of subject and object. To be sure, earlier Impressionist experiments had already produced a shift in attention to perceptual process. With Manet's *Bar*, this tendency was reinforced in a radical and decisive way.

From Manet, Vitz and Glimcher proceed to consider Paul Cézanne, observing that the work of the latter entails an even greater fractionation of classical perspective. Yet modernist art evolved in such a way that it did not merely challenge the classical outlook by introducing more and more ambiguity. In the transition from Manet and Cézanne to the Cubism of Braque and Picasso, there was an implicit move toward *resolving* the uncertainty. It is true that, in Picasso's *Card Player* (1913) and Braque's *Musical Forms* (1918) (Figs. 4a and 4b; see next page), ambiguous relationships can be observed. With respect to the Braque, for example, art theorist John Richardson notes that "'we can't settle upon the true relationships of the paper to the cardboard. Are the forms cut from the paper or are they the result of the cardboard overlaying it?'" (quoted in Vitz and Glimcher, pp.149–151). Yet the subtly disturbing quality of Manet's *Bar* clearly is absent from the more analytical-looking Cubist works. Gone is the sense of participating in a visual mystery. How can this "demystification" be explained?

Figure 4a. Pablo Picasso's *Card Player* (1913). Estate of Pablo Picasso (Paris)/ SODRAC (Montreal) 2003. Digital image © The Museum of Modern Art/Licensed by SCALA/Art Resource, NY.

Figure 4b. Georges Braque's *Musical Forms* (1918). Estate of Georges Braque / ADAGP (Paris) / SODRAC (Montreal) 2003. Philadelphia Museum of Art, Louise and Walter Arensberg Collection.

I venture to suggest that the ambiguous merging of subject and object implicit in Manet's *Bar* gives an intimation of the long-repressed *apeiron*. We saw in the last chapter that the rise of classical perspective brought an *idealization* of experience in which *apeiron*, while thereby being concealed, never ceased to constitute an underlying reality. It is from the undifferentiated matrix of *apeiron* that first comes *peras*: ratio, measure, differentiation, *perspective*. We found as well that the "rise of classical perspective" does not just refer to an event that took place hundreds of years ago. For the emergence of perspective from the pre-perspectival ground is reenacted early in the life of each contemporary individual following the "mirror stage" of his or her development. I now propose that—in addition to its phylogenetic and ontogenetic enactments—perspective arises "*micro*genetically" (Werner, 1948): With each and every momentary act of perception, a well-differentiated perspective crystallizes from the undifferentiated base of *apeiron* (see Rosen, 1999). These processes of perspectival differentiation are what classical idealization occludes in creating the impression that the fixed perspective always already exists. And it seems that Manet's *Bar*, in breaking the hold of classical perspective, provides its viewer with a glimpse of the pre-perspectival action of *apeiron* that first gives rise to the perspective s/he is viewing. It is in this specific sense that *Bar* calls attention to the *act of perception itself*, rather than being limited to an object that is perceived. As I view one perspective of *Bar*, I am haunted by the alter-perspective that hovers in the background. In this ambiguity of perspectives, I gain a glimmer of *apeiron*. The aura of mystery in the Manet, and in subsequent works by Cézanne, is thus indicative of resurgent *apeiron*. (We will return to Cézanne in Chapter Four.)

What Cubism seeks to do, for its part, is *banish* the specter of *apeiron*. This is achieved by exorcizing the ghostly presence of the alter-perspective. In Cubist works, no longer does that perspective hover in the background as we view the given perspective; rather, all perspectives are brought into the *foreground*, viewed simultaneously on the canvas. An act of abstraction is required to bring this about. The artist detaches himself more completely from the classical field of perception. As a result, while the forms may still appear to be ambiguous (as in Fig. 4), the artist himself no longer fully participates in this ambiguity. At the hands of the Cubists, the ambiguity that had been involved in *experiencing* the artwork is now captured on the canvas so as to become but an ambiguity in that which is *experienced*. In this process of abstraction whereby perspectival ambiguity is objectified, perspectives are more thoroughly flattened out. The instruments pictured in Braque's *Musical Forms*, for example, appear decidedly lacking in depth, are noticeably flatter than the more concretely representational figures portrayed in Manet's *Bar*. With the Braque, a more detached vantage point has been established from which we view with greater dispassion the ambiguity of the images; the tacit understanding is that, at this abstract level, ambiguity actually does not exist. Cubism therefore has been described as providing a "God's eye view" of reality (Pioch, 1995).

Equivalently, perhaps, Cubism gives us a "*mind's* eye view." Whereas Impressionism is primarily perceptual, Cubism is *conceptual*. To be sure, in viewing *Musical Forms*, we *perceive* ambiguity. But Braque does not want us to stop with that. We are expected to use our more abstract *conceptual* powers to fill the gaps left by the sensual eye, in that way bringing unity to our experience. No act of mere perception can reconcile the opposing perspectives we see, but, in the "*mind's* eye," they can be brought together as one. Thus, as Ernst H. Gombrich expressed it, Cubism was "the most radical attempt to stamp out ambiguity and to enforce one reading of the picture" (1960, p. 281).

We can understand better this effort to eliminate ambiguity when we realize that the project of Cubism is wholly analogous to that of Einstein. We have seen that Einstein's theory of relativity is essentially co-optative. It does accommodate *apeiron*, but in such a way that *apeiron* completely loses its force. This is essentially accomplished by accepting the dynamic ambiguity of nature on one level, while at once denying it on a level more basic. Even as Einstein appears to be supplanting the objectivity of classical physics with a relativistic account of nature's "observational events," he is in fact re-imposing fixed objectivity (see previous chapter). Cubism replaces the static objectivity of classical art with a "dynamic perceptual eventfulness" in the same fashion. In Einsteinian physics and Cubist art alike, the life of *apeiron* seems to dissipate in the rarefied climes of higher-dimensional modernist space. What I said about Cubism therefore can just as well be said of Einsteinian relativity: it does include perceptual process ("observational events"), but in a way in which all perspectival views (frames of reference or coordinate systems) are objectified, made explicit, rendered fully simultaneous "on the canvas"—that is, in the four-dimensional space-time continuum. In this regard, Gebser cites a work by Siegfried Giedion, which "is indebted to the thought of recent physics": Giedion "speaks of the 'dissolution of perspective' and the transformation of space into space-time evident in Cubism" (Gebser, 1985, p. 465).

Let me emphasize that, in the objective account the Cubist provides of subjective activity, he does not include his *own* subjectivity. Even when the Cubist paints a portrait of himself in an attempt at self-exposure that transgresses traditional boundaries, what he actually has done is turned himself—his dynamic internal process—into an object cast before the gaze of a new, more abstract subject tacitly operating behind the scenes. By no means is the latter subject exposed to view. In this way, modernism maintains the posture of object-in-space-before-subject.

Nevertheless, the modernist stance hardly has been unassailable—as our earlier examination of contemporary physics plainly demonstrates. We have seen that, as modern physics has gone beyond Einstein in pursuit of its long-held goal of unifying the diverse forces of nature, it has been brought to the brink of chaos, and that this has had the effect of *undermining* the posture of object-in-space-before-subject. It seems then that the presence of *apeiron* is very much in evidence on the contemporary scene, despite the bravado of those physicists who have promised that a "theory of everything" is just around the corner. In broader cultural terms, the subversion of the modernist project is termed *postmodernism*. How has postmodernism manifested itself in the realm of art?

During World War One, even while Cubism was still the dominant force in art, a pessimistic note was struck by a protest movement known as *Dadaism*. Repelled by the carnage of the war, Dadaists "decided to fight the 'scientific' and 'rational' war-makers and politicians with nonsense language, ugly dissonant music, and a totally irreverent and iconoclastic attitude toward the great masterpieces of past culture" (Cunningham and Reich, 1990, p. 371). Perhaps the best known Dada artist was Marcel Duchamp, whose mustachioed version of the *Mona Lisa* won him widespread notoriety. Far from seeking to reduce or eliminate uncertainty, Dadaism celebrated it with constructions such as collages of haphazardly assembled pieces of junk.

But modernism was not disposed of so easily. When the trauma of the First World War wore off a little, a new and more positive movement gained prominence in the artworld. *Surrealism* was surely influenced by Dada's radical critique of art, but it was shaped as well by the scientific psychology of Freud, and this was quintessentially modernist. The modernist project is implicit in the very term "psych-ology"—the application of *logos* to the *psyche*—and Surrealism was inspired by this effort to objectify the subject. Typically, the Surrealist surface is an eerie dreamscape of ordinary objects that are mysteriously distorted and/or oddly juxtaposed (this is well illustrated in what is probably the best known of Surrealist works: Salvador Dali's *The Persistence of Memory*). However, in contrast to the thoroughgoing irrationalism of Dada, the Surrealist departure from reason *is* but a surface phenomenon. Indeed, the Surrealist surface is like the manifest content of the Freudian dream. Probing beneath this veneer to the latent content, entering the depths of the unconscious, there is the promise of meaning and order.

The influence of Surrealism on twentieth century art was considerable. It was Surrealism that gave Abstract Expressionism its impetus to tap the unconscious through spontaneous action with canvas and sculpture. And

Abstract Expressionism, in its turn, profoundly affected the artistic movements that succeeded *it*. However, following World War Two, art was gradually transformed in an important way: The original spirit of Dada returned, this time to stay. It was the devastation of World War One that had brought Dadaism's initial flirtation with postmodernity. After the rather more horrific Second World War—the war that culminated in Auschwitz and Hiroshima, the postmodern self-deconstruction of art began to come into its own. We see this particularly in Jasper Johns's and Andy Warhol's irreverent send up of "high art" by celebrating the banalities of popular culture. Following the rise of Pop Art, a diverse assortment of other strategies and styles began to proliferate: the absurdist "happenings" of Performance Art, the sometimes-apocalyptic emotionality of Neo-Expressionism, and so on. As the artworld has unraveled over the last quarter century into a pluralistic hodgepodge of divergent approaches, the only threads of unity that have remained appear to be negative ones: the continuing emphasis on uncertainty, fragmentation, and eccentric absurdity; the agreement that art can never be fully defined; the effort by art to call into question its own existence. We can summarize the crux of the difference between the now-prevalent postmodern art and its modernist predecessor in this way: Whereas the latter sought unitary meaning beneath the surface play of ambiguities, for the former, there is *only* that surface.

Perhaps the foremost ambiguity of postmodernism is that between subject and object, self and other. No longer is art simply a question of expressing oneself freely or "finding" oneself, since the very notion of the self is problematic for the postmodern artist. With the shattering of modernism's "vision of unity," the position the modernist subject thought he had secured for himself is now recognized as highly precarious. An identity crisis of unprecedented magnitude has ensued. In short, postmodern art intimates the collapse of rational order into the chaos of *apeiron*. It is not so much that *apeiron* is recognized and acknowledged as such: as *prime matter*, the tempestuous goddess, arational Mother Nature. Rather, postmodern art is essentially a negative enterprise. It negates modernism without offering a viable alternative; as a consequence, modernism—instead of being surpassed—lingers on in its own wreckage. Behind the addictively obsessive quality of postmodern art—exemplified by the endless repetition of a single image in the works of Andy Warhol and others—lies the inability to end the modernist quest for abstract unity, even though it is now obvious that the quest has failed. The contemporary artworld thus confronts its own "black hole."

For my second example of how developments in science are reflected in the broader cultural context, I turn to the field of *media*.

We have been considering the return of *apeiron* in the Age of Light, modernism's attempt to circumvent *apeiron*, and the collapse of this effort with twentieth century postmodernism. I propose that this very same historical sequence is reflected in three advances in technology: the photograph, cinema, and television, respectively. (In the next chapter, we will discuss the computer.)

According to media theorist Marshall McLuhan, "Photography was ...decisive in making the break between mere mechanical industrialism and the graphic age of electronic man. The step from the age of Typographic Man [the era of the printed word inaugurated by Gutenberg] to the age of Graphic Man was taken with the invention of photography" (1964, p. 171). It was just around the time when Manet was preparing to overthrow perspective in the field of art that the new technique of photography was burgeoning, having been invented in 1839. Before then, if one sought to capture one's perceptions of the external, three-dimensionally experienced world on a two-dimensional surface, the process would "take time." That is to say, the act of recreating objective reality necessarily would unfold over a substantial duration. With the photograph, the subject's reproduction of the object literally occurs in a flash, in the fleeting instant required for the capturing of light (electromagnetic energy). The subjective time sense of the photographer therefore is markedly different from that of the classical artist. True, the serious photographer may require an extensive period to *set up* a photograph. But the act of taking the photograph, considered in itself, implies a sense of time that was entirely new in the middle of the nineteenth century. The continuous flow from the past into the future that had characterized earlier temporal experience was ruptured by a decontextualizing perception of instantaneous leaping. Thus McLuhan could say that, with photography, "the effect of speeding up temporal sequence is to abolish time" (1964, p. 176).

Berger and Mohr (1982) elaborate on this, demonstrating that photography brings a sense of temporal discontinuity not only to the reproductive process, but to the resulting product as well (just as artistic works created in classical time bear the mark of past-to-future flow in their texture, non-uniformity, etc.):

> A photograph arrests the flow of time in which the event photographed once existed. All photographs are of the past [like all drawings and paintings], yet in them an instant of the past is arrested so that, unlike a lived past, it can never lead to the present. Every photograph presents us with two messages: a message concerning the event photographed and another concerning a shock of discontinuity. [1982, p. 86]....An instant photographed can only acquire meaning insofar as the viewer can

> read into it a duration extending beyond itself. When we find a photograph meaningful, we are lending it a past and a future. (1982, p. 89)

In short, the non-mechanical, instantaneous flash that produces the "snapshot" (the chemically fixed record of an electromagnetic event) shatters the classical sense of temporal duration.

In comparing photographic time with the temporal character of drawing, Berger and Mohr (1982) note the relative *homogeneity* of the former:

> The time which exists within a drawing is not uniform. The artist gives more time to what she or he considers important. A face is likely to contain more time than the sky above it....In a photograph time is uniform: every part of the image has been subjected to a chemical process of uniform duration. (1982, p. 95)

This loss of temporal contrast suggests a process of leveling wherein classical meaning and quality, which *depend* on textured contrasts and gradations, are flattened out, neutralized by indifferent equalization. In the visual medium of the photograph, nothing has more significance than anything else, not even the human face; therefore, nothing has significance—except what we *import*. Like art and physics in the Age of Light, there is indeed a flatness to the photograph indicative of the collapse of classical perspective and the whole system of values with which it was integrally associated.

Berger and Mohr go on to emphasize a closely related characteristic of the photograph: "All photographs are ambiguous. All photographs have been taken out of a continuity....Discontinuity always produces ambiguity" (1982, p. 91). We have seen how ambiguity arose in nineteenth century science and art alike as a result of fracturing the perspectival unity of classical space. We know as well that, in both cases, the ambiguity was resisted, that an effort was made to dispel it, to restore unity to experience by resorting to higher levels of abstraction. Both Cubist and Einsteinian space-time—which in fact were unveiled at the same historical moment (the first decade of the twentieth century)—served essentially as modernist surrogates for the classical unity that had been lost. What may we say of the *photographic* ambiguity that arose in the middle of the nineteenth century? Did it occasion a distinct modernist response analogous to that given in the fields of art and theoretical science? It most certainly did. Photography was succeeded by the *cinema*.

The chronological parallelism is noteworthy. In the last two decades of the nineteenth century, just when Einstein was incubating his response to the question raised by Michelson and Morley and art was entering its proto-Cubist phase, the motion picture was emerging as a brand new medium. In keeping with the modernist character of Cubism and Einsteinian physics, cinematic time aimed—in its abstract way, at restoring the continuity of concrete time that had been broken by the advent of photography. In the process, ambiguity was to be eliminated. McLuhan thus related the beginnings of the cinema to Gombrich's assessment of Cubism (cited above) in the following way: "[It] was at this moment of the movie that Cubism occurred, and it [Cubism] has been described by E. H. Gombrich...as 'the most radical attempt to stamp out ambiguity and to enforce one reading of the picture—that of a man-made construction, a colored canvas'" (1964, p. 27). In what specific way did the movie, in its own right, seek to stamp out ambiguity? The basic principle of cinema is simple and well known: an event is recorded by a series of still shots, each member of which, considered by itself, entails the photographic ambiguity. The shots are then exposed to the viewer in a mechanical sequence that is rapid enough to convey the impression of motion, a time-directed movement from past to future that reconstructs the event within a durational context in which photographic ambiguity appears to dissolve. The "shock of discontinuity" spoken of by Berger and Mohr is banished and a sense of temporal flow is restored.

McLuhan portrays cinema as the heir to typography, stressing the similarity between the two media. Gutenberg type is the "hot" medium that spearheaded the break with medieval culture. The well-differentiated, standardized, sequentially processed units of type were tailor-made for the segmented and sequential operations of the high-profile, sharply individualized Renaissance man, the ego who began asserting himself aggressively in the new world of mechanical causes and effects unfolding in linear time. ("Literate [typographic] people think of cause and effect as sequential," comments McLuhan, "as if one thing pushed another along by physical force"; 1964, p. 251.) When the mechanical order was challenged by the instantaneity of electromagnetism, the response was the invention of a new "hot" medium, the hybrid medium of the movie, described by McLuhan as "a spectacular wedding of the old mechanical technology and the new electric world" (1964, p. 249). McLuhan contends that film technology did much to "continue" and "surpass" the purely mechanical book technology (p. 255), that it "pushed this mechanism to the utmost mechanical verge and beyond" (p. 254), that cinema "was, as a form, the final fulfillment of the great potential of typographic fragmentation" (p. 257). In these passages, no accuracy would be lost if, for "film technology," we read "Einsteinian relativity" or "Cubist art," and for "book technology," we substituted "Newtonian physics" or "classical art." The fundamental basis for the success of cinema is essentially the same as that for relativity theory or Cubism: while effectively accommodating itself to the electromagnetic revolution on one level, on another it managed to uphold the classical sense of reality, to preserve a semblance of the post-Renaissance ego's way of experiencing space and time. Note that, in the case of relativity theory, even though time is mathematically unified with space, the past and future are partitioned from each other in such a way that time actually retains its classical character of simple successiveness (see Griffin, 1986, for some illuminating discourse on the nature of Einsteinian time; see also, Rosen, 1994, Chapter Five).

But I must stress that Einsteinian time is by no means *simply* classical. Instead, it is an *abstraction* upon the pre-electronic awareness of time that is one step removed from it. By the same token, while the cinematic medium maintains duration amidst electromagnetic instantaneity, it does so in an altered, abstracted form. Behind the simulated appearance of durational flow in the "reel world" (as McLuhan ironically calls it) is but a compilation of frozen instants, a discontinuous series of durationless "nows." So it is not surprising that, to this day, we refer to the cinema as "the flicks": at bottom the movie is "a jerky mechanical ballet of flicks" (McLuhan, 1964, p. 254).

(Given the close relationship between cinema and Cubism, we may say that, in both cases, the problem to be solved is basically that of "flicker fusion." If the photographic stills can be presented to the viewer at an appropriately rapid rate, the viewer will be able to bridge the perceptual gap between the separate frames via an act of *con*ceptual integration. When the "mind's eye" thus succeeds in abstractly filling the gaps left by the sensual eye, the flicker becomes subliminal and the appearance of a seamless temporal flow is achieved. Just such a fusion of perspectives is required of the viewer of the Cubist work. Therefore, assessed from the modernist standpoint that places a premium on the abstract unity of experience, "proto-Cubist" artworks such as Manet's *Bar* may be likened to early cinematic efforts in which the "flicker" is still quite noticeable.)

Most essentially, the gap in the continuum that modernism sought to fill was a gap in human being itself. The Age of Light was indeed a watershed in human development. What we saw in the last chapter is that the long history of space and time reflects a process in which the individual emerged from the chaotic womb of nature, forging a distinctive sense of identity and personal agency. In the course of thus detaching himself from *apeiron*, the upsurgent ego turned *apeiron* into an object, thereby obscuring her true nature. Then came the nineteenth century, a time when the process of individuation, in approaching its climax, gave a new intimation of *apeiron*, one that appeared to threaten the very life of the *cogito*. Small wonder that the ego would resort to compensatory measures. Hence the cinema, a world of "canned dreams...an inward world of fantasy [in which the viewer]...sits in psychological solitude" (McLuhan, 1964, p. 255), "a sheer dream world of romantic illusions" (p. 254). Since the ego had long ago ceased to participate integrally in nature herself, Hollywood offered the modernist ego the only kind of substitute it seemed prepared to accept for its lost integrity: egoistic dreams of triumph in an objectified natural world; an endless variety of "success stories"; countless tales of egos on the ascent with whom viewers vicariously could identify—egos that, if not transcendent in victory, were transcendent in defeat. Whatever the specific content of these films, the fundamental "triumph" of the egos they portrayed lay in the magical metamorphosis inherent to the medium: the modernist transformation of the finitized, mortalized, hole-ridden nineteenth century individual into an abstract image of infinity, immortality, and wholeness. "The medium is the message" says McLuhan, and the essential message of the cinematic medium is that continuity can arise from discontinuity "without a flicker."

Nevertheless, the modernist message of the cinema eventually began to wear thin. If it is reasonable to say of the modernist era in general what McLuhan said of cinema—that it was a "hot," "explosive" epoch in human development—then it might also be said that the climactic "explosion" of modernism occurred with two unprecedented catastrophes: the incineration of millions by the Nazis during World War Two and the dropping of the atomic bomb on Japan at the end of that war. Ever since Auschwitz and Hiroshima gave their testimony to the folly of the modernist ego, human affairs have been "cooling down" considerably. The ego has been "imploding," McLuhan would say. Exhausted and disillusioned, it has been succumbing to the mood of *post*modernism. And this has been accompanied by the proliferation of a new medium eclipsing the cinema and designated "cool" by McLuhan. I am speaking of television.

TV's basic operating principles are not difficult to describe. The television camera contains a photosensitive surface upon which light reflected from the televised event continuously falls. Unlike the photographic plate wherein light waves are chemically fixed, the photosensitive TV plate is rapidly scanned by a moving beam of particles shot from an electron gun. Light waves are thus broken down into discrete electronic impulses, which are then transmitted over a distance via radio waves (broadcast TV) or wire (cable TV). At the receiving end, the electronic impulses are projected onto a luminescent screen upon which the original images are reconstructed in the process of viewing. So watching TV is an act of reconstructing deconstructed images from moment to moment, rendering continuous that which, an instant before, had been rendered discrete; each instant, the viewer must "put humpty dumpty back together again." Or in the words of McLuhan, "The TV image is not a *still* shot. It is not a photo…but a ceaselessly forming contour of things limned by the scanning-finger" (1964, p. 272).

To bring out the essential character of television, let me refine what I said above about photography. The product that results from trapping light waves in a chemical emulsion surely is flattened, uniform, time-discontinuous, and thus ambiguous—relative to a perspectival drawing or painting (as

Berger and Mohr observed). Yet, compared with the television image, the leveling of perspective and interruption of time involved in photography is actually incomplete. This distinction is implicit in McLuhan's comment that, whereas "film and photo...have a built-in angle of vision...the third dimension is alien to TV....[T]he TV image...is a [completely] flat two-dimensional mosaic" (1964, p. 273). Why does photography not flatten perspective and freeze the flow of time as completely as TV? Although the introduction of the former technique did mark the transition from the mechanical to the electronic age, photography is not *purely* electronic but is a process that does also involve chemistry. As a chemical recording of light waves, the photographic flash is a supra-atomic event entailing the molecular bonding of atoms. For perspectival collapse to be complete, for a *total* loss of temporal continuity, what is required is the quantized, utterly discontinuous leaping of sub-atomic particles—as in fact is found in TV. Since photography stops short of this, a modicum of information is retained that is relevant to temporal flow. When light waves are fixed by chemicals and therefore not reduced to the informationless uniformity of microscopic particles, they preserve macroscopic relationships that permit some inferences to be made concerning angle, distance, and depth. More specifically, the photograph maintains a modest sense of the differences in the distances over which light was reflected from the various planes of the photographed object, and this conveys a sense of nonsimultaneity, temporal differentiation—though a far weaker one than in the drawing or painting, to be sure.

It is in television, the purely electronic medium, that the freezing of classical time reaches a consummation. TV is thus a most ambiguous medium, one that "offers little detail and a low degree of information, much like the cartoon" (McLuhan, 1964, p. 273). Indeed, television is the medium through which the ambiguity of *apeiron*, having been temporarily held at bay by the cinema, now returns with a vengeance. With television, the gradations of classical experience—the textures, qualities, nuances, and meanings—are all completely degraded. Quality is replaced by sheer quantity, empty repetition (compare this with the *art* of the time, and, in particular, the works of Andy Warhol). This is the "message" behind the incessant electronic jumping of the "scanning-finger" from dot to dot across the luminescent screen. Here the mood of postmodernism is quintessentially expressed.

The "message" was not as clear in the early days of television as it presently is. Superficially at least, TV can, after all, function as a modernist medium; it can play the cinematic role of creating the illusion of durational flow (though the discerning eye will recognize that it does so in a strained, cartoon-like fashion). In its formative period, under the lingering influence of the medium it was succeeding, television did effect a "filmic" approach, miming the pseudo-continuities of the silver screen. Today this is no longer true. Today television is "coming into its own." What this means is that the medium is becoming the message in a fully-fledged, wholly transparent way, that the content and form of the TV images presented to the viewer are coming to fully and openly reflect the "deep structure" of this medium, the underlying principle of radical discontinuity upon which it operates.

The fragmented pulse of contemporary television is so pervasive it hardly needs to be documented. The subliminal leaping of electrons constituting the fundamental event of the medium has been raised above the threshold of awareness, mirrored in the frenzied kaleidoscope of juxtaposed, overlapped images flooding the senses of viewers in too many programs to name. With abstracted movement as such gaining precedence over the concretely meaningful contents among which the movement occurs, ambiguity reaches a maximum. Consider a prime example of this: the programming of MTV (the Music Television Network), which, in fact, is explicitly headlined as "Postmodern Television." The bulk of the fare on MTV consists of music videos, discrete visual pieces of brief duration (normally not more than several minutes), each of which itself is highly fragmented into shots enduring for fractions of seconds. Little coherent relationship exists between these disjointed images—at least not by the old, literate standards of coherence that depend on syntactical continuity and discursive meaning. It is programming of this kind that prompted O. B. Hardison (1989, p. 178) to comment that television generally makes the discontinuities of Surrealist filmmakers like Jean Cocteau seem tame and old-fashioned.

Now, it is a well-known fact that music videos can have an especially "hypnotic" effect on their viewers, riveting them to their TV receivers in spellbound rapture for hours on end. This might be understood in terms of the following analogy. Physiological research suggests that if the human brain is irradiated with energy pulses whose frequencies are synchronous with that of the brain itself, a certain resonance is established and the brain can be "entrained" to the incoming energy; that is, a trance-like state of attunement to it can be induced (see Jantsch, 1980, pp. 213–214). Is a similar effect involved in the attunement of young people to the pulses of MTV? Are habitual viewers of music videos "hypnotized" by what they see because the consciously registering rapid pulsations of the televised images are highly synchronized with the subliminal pulsation of the electronic scanning-finger? Is this what happens when the "message" of the medium is fully conveyed? To provide a further illustration of the possibility that television viewing can be "hypnotic" when a sufficient level of synchrony is attained, let me describe a certain change that took place in my personal viewing behavior. Perhaps the phenomenology of this anecdote will resonate with your own experience.

It was not that long ago that one facet of my experience with TV did remain mechanical: To change channels, I had to manually crank the tuner. Then I switched to cable TV and was provided with an electronic device for effectively navigating the greatly expanded array of program possibilities. Granted this "magical power" of instantaneous access, my viewing behavior underwent a change. Often, in watching a given program, I would find myself soon becoming bored and restless, and would feel inclined to press the buttons on my remote control to see what was happening elsewhere. This lowered threshold of boredom and increased tendency to "channel hop" persists in my present viewing habits. In fact, at times I find myself disposed to jump from channel to channel purely for the sake of the jumping. When I indulge myself in such activity for a long enough spell, I seem to enter a trancelike state of vacuous drift, an inert frame of mind wherein the contents of the particular programs "flatten out" so completely they lose all interest and meaning. This, I submit, is postmodern behavior *par excellence*. Engaging in such "somnambulistic" action, am I not in high synchrony with the fundamental action of TV's electronic scanning-finger? In this desultory state, do I not unwittingly mime TV in its essence, write its underlying message with my very own body? Again, the message is that of quantity utterly stripped of quality, of pure homogeneity, the repetition of repetition itself and *only* repetition. It is not so much an ambiguity of meaning that I experience here, for ambiguity has now become so pervasive that there *is no* meaning.

In examining postmodern art, I intimated that the obsessive fixation upon a single image found in many of the works has an addictive quality. The experience of watching television makes it even clearer that the *modus vivendi* of the postmodern ego is that of addiction. Since the underlying message of the TV medium is that of utter ambiguity and emptiness, what is really mirrored back to me on the screen is not the positive presence of others that confirms and validates my own presence; rather, in gazing into the tube, I see my own negation. Therefore, to encounter the "black hole," I need not embark on an excursion to the stars: It is right there in front of me every time I switch on the power. How do I react to this? By what McLuhan calls "involvement in depth" (1964, p. 272).

In viewing TV, I enter into the emptiness before me in a futile attempt to fill the vacuum by drawing from the hollow recesses of the being whose image is reflected. Necessarily, all I come up with in this vain effort is more emptiness. The postmodern ego's compulsive attempt to fill itself from its own emptiness naturally fails, and, left entirely unsatisfied, it will try again and again in a viciously self-defeating cycle of addiction. (If we find cinematic experience more satisfying than TV, it is because the surrogates for life offered by the former are more substantial and convincing than the cartoon-like, paper-thin images of the latter.) So I am "watching TV more these days, and enjoying it less" (to echo an old TV slogan that actually was directed to the habit of smoking), switching channels more and gaining little satisfaction, until I reach the point where there is no satisfaction whatsoever. Yet I continue to watch, for TV is my mirror and where else does one look to confirm one's identity but in the mirror? Do I not see my own negation each time I look? Then I must look again. (Note that McLuhan himself put a more positive spin on the TV viewer's "involvement in depth," seeing it as indicating participation in the emerging "global village." This reading is countered by communication theorist Walter Ong's observation that the "television age," the "age of 'togetherness'...is also the age of the isolated individual" [1977, p. 341].)

The TV experience of empty repetition brings to mind the examples given in the previous chapter of the way in which basic change is resisted when a strong commitment has been made to maintaining the status quo. When the ancient Greek order of perfect circularity was called into question by certain astronomical observations, instead of relinquishing the idea of the circle, its image was gratuitously reproduced in the Ptolemaic epicycles, despite the fact that the circle had actually lost its original meaning. Similarly, when the fundamental concept of continuity was challenged by the phenomena of quantum mechanics, the response was to replicate the continuum ad infinitum in the subspaces of Hilbert space. I submit that this tendency to obsess upon an idea or image, to invoke it again and again—as if its loss of meaning could be compensated for by its mass production, as if quantity could take the place of quality—is what addiction is all about. Thus the "mass production" of continuity in postmodern physics is reflected in the productions of postmodern art and television, and in "market-oriented" postmodern culture in general. More and more, we try to fill the "black hole" in ourselves with MORE, but succeed only in deepening that hole.

The black hole is a "cool" phenomenon, as McLuhan would say. It is the implosion that follows the explosive phase of a star's history, when the impetuous heat of its "youth" has given way to more sober gravitational concerns. Speaking in broader cultural terms, McLuhan is telling us that it is the *human ego* that has imploded. Since the end of World War Two, the limitations of the ego have certainly become more difficult to deny, and human experience has changed accordingly. Yet, if phenomena like Pop Art, the "anti-hero," and "rap" express a new mood of iconoclasm that reflects the ego's recognition of its own finitude, the hole at its center—at the very same time, the ego also still seeks "coolly" to *deny* that hole. This sums up the essence of the addictive age of television: an implicit recognition of our emptiness that simultaneously strives to deny it. Yes, we sense the hole more clearly now than the modernist did; the hole at the center of our being has become *so* transparent that it scarcely can be ignored. Yet, "playing it cool," we nevertheless continue our denial, but now in such a way that its very quality of ineffectual repetition actually accentuates the gnawing emptiness at the core of us.

The black hole is indeed a "cool" phenomenon, at least when viewed from the "outside," that is, from the standpoint of the life of the star, whose extroverted energies have now dissipated. We know, however, that the *interior* of the hole is hardly lacking in energy, albeit of an introverted, implosive sort. The further we go into the heart of the black hole, the greater this crush of gravity becomes, until a marked turbulence sets in. The dynamic flux in question is not that of an event taking place in space and time; rather, it is the flux of *space and time themselves*. What this means is that spatiotemporal discontinuity attends the movement to the singular core of the black hole; concomitantly, there is ambiguity, all-pervasive uncertainty. When the core is finally reached, the chaos no longer can be contained. The infinite gravitational energy confronted at the center of the black hole is of course the energy of *apeiron*.

It is *apeiron* that the *ego* confronts with the postmodern implosion of Western culture, and *apeiron* that it seeks most desperately to deny. Long has this "Black Goddess" (Woodman, 1996) been repressed, for she threatens the very life of the rational being. The existence of *homo rationalis* is predicated on an assumption that itself has been obscured for centuries. Never enunciated but always taken for granted is the supposition that "man" has contained "Mother Nature" and now stands free and clear of her. It is this posture of detachment and power of containment that most essentially defines the individual. He maintains himself *as* an in-dividual—as he who is one, undivided, unitary—by containing *her*. At bottom, this is what science's quest for unity has been all about.

Plato's "leaky vessel" had been a tentative step toward containment. Descartes and Newton did far better. With them, *apeiron* was so well bottled up that there was not so much as a hint of her. But then, in the middle of the nineteenth century, just when the work of science seemed on the verge of completion, just when Mother Nature was to be fully contained once and for all, the container gave evidence of fissures. For a hundred years thereafter, we did our modernist best to plug the leaks, to keep the genie in the bottle—until the fateful morning of 6 August 1945. *That* morning more than any other, with its frightful unleashing of mushroom-shaped chaos, perhaps best symbolizes the failure of containment. Yet this *flattening* of an entire city was only the *beginning* of the end, for the thermonuclear outpouring of energy in fact *was* contained, limited as it was to a relatively small geographical area. The tidal wave of energy set free by opening up the atom is but a droplet in the ocean of energy that seethes beneath the *sub-atomic limit* of the microworld, beneath the Planck length. This is the "zero-point" energy, the energy of the black hole, the unbridled fury of the "Black Goddess." If *that* energy were set loose, the container that would be exploded would not be confinable to a region of space since it would be *space itself* that would explode. Where then would the energy "go"?

Containment and detachment are integral aspects of the same process: by the same act in which the subject seals its object into space, its seals itself out, separates itself from that object. It is clear then where the primal energy would have to flow were the spatial seal to be broken. Shattering the spatial container means *ending the division of subject and object*; therefore, the energy that had been bottled up in containing *apeiron* as but an object would now overflow the ontological boundary, flow freely back into the subject. The ambiguous blending of subject and object intimated earlier would now be realized in full. With this "flooding" of the subject, the world as we have come to know it—that governed by the ancient formula, *object-in-space-before-(detached)-subject*—would come to an end.

No wonder the specter of *apeiron* is so fearsome to behold. It is not just the lives of particular individuals that seem to be at risk here, but the overall process of individuation that has marked the progress of humanity for thousands of years. Then does the coming to presence of the black hole necessarily foreshadow the wholesale regression of our species? Does the vision of the "Black Goddess" betoken the unavoidable demise of rational thinking, of language and reflective individuality? Are these gifts that define the unique heritage of humankind simply to be lost? In concluding Chapter One, I hinted that *apeiron*—far from merely signaling the end of individuation—is the key to its genuine completion. In the chapters to follow, I intend to make this clear. For now, I will close the present chapter with an alchemical preview: the story of the "Spirit in the Bottle."

According to C. G. Jung, this Grimms' fairy tale "contains the quintessence and deepest meaning of the Hermetic Mystery": a powerful spirit is trapped in the earth beneath an oak tree, enclosed within "a well-sealed glass bottle" (1967, p. 193). Hearing the spirit cry "Let me out!" a passing youth opens the bottle, whereupon the spirit rushes forth, identifies itself as mighty Mercurius, and threatens to strangle its liberator. But the boy tricks the spirit back into the bottle, and the tamed Mercurius then promises that, if freed again, it will serve the boy in a beneficial way.

Jung relates the glass bottle of the fairy tale to the Hermetic vessel of alchemy, the container in which the transformation of *prima materia* was supposed to take place:

> The bottle is an artificial human product and thus signifies the intellectual purposefulness and artificiality of the [alchemical] procedure,

> whose obvious aim is to isolate the spirit from the surrounding medium. As the *vas Hermeticum* of alchemy, it was "hermetically" sealed....[I]t had to be made of glass, and had also to be as round as possible, since it was meant to represent the cosmos in which the earth was created. (Jung 1967, p. 197)

In Jung's interpretation, the Mercurial spirit represented to the alchemists the initially unconscious, wildly irrational power of instinct, of embodied nature—of *apeiron*. In turn, the "bottling up" of Mercurius signified the necessity of gaining intellectual control over nature. Jung took the fashioning of the Hermetic vessel to symbolize the process of individuation that culminates in the *unio mentalis* (1970, p. 465), a state of intellectual maturity that brings mental development to fruition. This is the challenge symbolically faced by the boy in the Grimms' fairy tale, and the challenge the alchemist faced. Prior to "properly sealing the bottle," Mercurius was always a threat to escape, a regression to the primal past that would overwhelm the alchemist. In the meanwhile, the alchemist had to keep the "spirit" imprisoned as best he could.

Alchemy actually did take some major strides toward effectively containing Mercurius. How was this done? It was carried out through the revolution in thinking that occurred with the Renaissance, and the attendant rise of scientific rationalism and modern mathematics. The *Cartesian* vessel—that is, perspectival space—indeed contained "Mercurius" better than any of its predecessors had. But doesn't modern science *repudiate* alchemy? In actuality, the first modern scientists were themselves alchemists (Berman, 1989). Seen from this point of view, the scientific enterprise may be regarded as having originated in the goal of furthering alchemy's program of individuation, not renouncing it.

Nevertheless, it is clear from our examination of modern physics and other contemporary spheres of endeavor that, after four hundred years, the vessel has still not been sealed hermetically, and that now, more than ever before, "Mercurius" is straining to break free. Then what would it actually mean to properly seal the vessel? If "Mercurius" is in fact irrepressible, it could not simply be a matter of finding a more effective way of keeping it inside the bottle. And yet, the vessel is to be hermetically sealed. Herein lies the ultimate paradox of individuation: How can *apeiron* be fully contained, and, at the same time, fully liberated? And this is the ultimate question in the pages that follow.

Part Two

COAGULA
THE RETURN TO *APEIRON*

Three

PHILOSOPHICAL PRECURSORS

1. The "Other Copenhagen"

The Danish physicist Niels Bohr was a principal author of the Copenhagen Interpretation of quantum mechanics. According to Gerald Holton (1988), Bohr was significantly influenced by another prominent resident of Copenhagen: the existentialist philosopher Søren Kierkegaard (1813–1855).

Holton recounts how,

> [I]n 1909 Niels sent his brother Harald as a birthday gift Kierkegaard's book, *Stages on Life's Way*, with a letter saying, "It is the only thing I have to send; but I do not believe that it would be very easy to find anything better. In any case I have had very much pleasure in reading it, I even think that it is one of the most delightful things I have ever read." (1988, pp. 128–129)

In Holton's opinion, Bohr did not simply enjoy reading Kierkegaard but was seriously affected by what he read. This view gains credence when we recall three key features of quantum mechanics identified in Part One: discontinuity, uncertainty, and subject-object interrelatedness. These very same elements play a pivotal role in the philosophy of Kierkegaard.

Kierkegaard passionately dedicated himself to attacking the dispassionate posture of classical science, and of post-Renaissance culture in general. The classical subject abstracts himself from the world, detaches himself from its earthly affairs, and, in so doing, renders himself anonymous. In adopting this impersonal stance, he is answerable to no one. This creates an ethical dilemma: "the impersonal, which for the most part is irresponsible ...is essentially demoralizing....[A]nonymity, as the most absolute expression for the impersonal, the irresponsible...is a fundamental source of the modern demoralization" (1848/1975b, p. 88). Moreover, in removing oneself from concrete involvement in the world, one removes oneself from *the truth*.

"The crowd is untruth," proclaims Kierkegaard (1859/1975, p. 96). "[N]o witness for the truth dare become engaged with the crowd" (p. 97). For, in the anonymity of the "crowd"—which may be an actual assemblage of people, or, what is far more anonymous, an abstract institution such as science—one *loses oneself*; in so doing, one loses contact with the truth.

Kierkegaard would say that we could not subtract from the truth the actuality of the self, the existential reality of the concrete individual, and still have the truth. Yet this is precisely what science seeks to do. Making the point a little differently, in his *Concluding Unscientific Postscript* Kierkegaard asserts that "Truth is Subjectivity" (1846/1975, p. 110). Whereas science's "*objective accent falls on WHAT is said, the subjective accent [is] on HOW it is said*....[H]ow refers to the relationship sustained by the existing individual, in his own existence, to the content of his utterance. Objectively the interest is focused merely on the thought-content, subjectively on the inwardness" (1846/1975, p. 116). Is Kierkegaard calling here simply for a reversal of the scientist's tendency to direct his attention exclusively to the object in concealment of his own subjectivity? Are we now to focus *only* on the subject and *neglect* the object? Evidently not. What Kierkegaard wants us to do is pay attention to the *relationship* between subject and object, between the "existing individual" and the "content of his utterance." So it is not so much that we are to gain access to the truth simply by shifting our awareness from object to subject, but by an act of concrete self-reference in which awareness expands from science's hitherto narrow preoccupation with the object alone and now includes the subject *in relation to* that object. In other words, the subject, instead of remaining a shadowy figure in the background of awareness, enters the foreground in such a way that its intimate relationship to its object is now disclosed. Of course, the change of focus urged by Kierkegaard would constitute a fundamental challenge to the classical presupposition of object-in-space-before-(anonymous-)subject.

Since the old "objective truth" was rooted in the categorial division of subject and object, Kierkegaard's radical questioning of that division was bound to result in a profound sense of uncertainty. Kierkegaard was well aware of this. In fact, what is required, he said, is that one "embraces [the] uncertainty" (p. 115). Truth itself is defined in these terms: "*An objective uncertainty held fast in an appropriation-process of the most passionate inwardness is the truth*" (p. 117). For Kierkegaard then, "truth is precisely the venture which chooses an objective uncertainty with the passion of the infinite" (p. 117). Moreover, an even greater truth can be realized if one embraces *paradox*:

> [T]he more objective security the less inwardness (for inwardness is precisely subjectivity), and the less objective security the more profound the possible inwardness. When the paradox is paradoxical in itself, it repels the individual by virtue of its absurdity, and the corresponding passion of inwardness is...the truth. (1846/1975, p. 117)

Evidently then, mere uncertainty about what is taken as objective does not go far enough, since, in simply raising doubts about objective truth, one still leaves open the possibility that those doubts might be dispelled. It is when doubt becomes absurdity, contradiction, that there is no longer any doubt about the inadequacy of objectivism.

Now, Kierkegaard disparaged the attempt to approach the truth by the gradual "approximation-process" (p. 118) employed in objective scientific inquiry. Our realization of the truth is not something we can assemble in small increments, bit by bit. There must be a *leap of faith* in which the truth is known as a whole. Why is it that science, for its part, is obliged to proceed in a step-by-step, piecemeal fashion? This is an inevitable consequence of splitting subject and object in such a way that the truth becomes narrowed to the object-*in-space*. We saw in Part One that classical science's spatial container is an infinitely divisible continuum. The objects contained in this medium must themselves be susceptible to unlimited division. The complete decomposability of the object implies that its elements are related to one another in a strictly external manner. Therefore, rather than constituting a genuinely cohesive, *in*divisible whole, the object-in-space can be nothing more than a superficial assemblage of parts. Given the absence of cohesive wholeness, the parts must take precedence. In any investigation of an object-in-space, our analysis must *begin* with the parts, which we then hope to piece together so as to *approximate* the "whole." But, as Kierkegaard says, the process is "never-ending" (1846/1975, p. 115), since parts that are related in a merely external way can never add up to an authentically integrated whole. What is the missing element of cohesion that prevents the object-in-space from entailing a whole truth? Kierkegaard would say it is the *subject*, and we can see why this is so. For it is by first dividing the object from the subject that truth is narrowed into an object-in-space that is itself infinitely divisible. Thus, behind the limitless divisibility of the object that precludes its cohesion, is the division of object and subject. This subject, in donning his cloak of anonymity, is subtracted from the truth, which leaves but a partial truth.

Then, for the whole truth to be realized, the subject must remove his cloak, expose himself to the world, become a full-fledged participant in that world. Thus exposed, how would this subject appear to a world that had never known his likes? Evidently, he would be manifested as uncertainty—no, as contradiction, absurdity. In unveiling the subject, a *hole* would be created in the continuum of classical science. The piecemeal truth of the continuum would be shaken by a *dis*continuous verity, one requiring the Kierkegaardian "leap of faith." What we are seeing here is the same relationship between discontinuity and subject-object interrelatedness that we encountered in Part One: To bring subject and object together as Kierkegaard exhorts would be to break the seal on the continuum that had been forged by the old Cartesian subject in detaching himself from his object-in-space.

Kierkegaard saw the realization of lived subjectivity in *developmental* terms. The central question for him was that of *individuation*: "'The Individual' is the category through which…this age, all history, the human race as a whole, must pass" (1859/1975, p. 100). Yet Kierkegaard takes note of the difficulty of embracing one's individuality, for "this view involves an admission of weakness and impotence, and seems therefore far from inviting, and for this reason perhaps is…seldom heard" (p. 99). Thus, rather than facing the daunting challenge of becoming an individual, people are more likely to take refuge in the abstractions of the crowd, where—striking a pose of invulnerability—they can indulge themselves in the fantasy of being masters of the world without ever needing to put themselves at risk. Submerged in the crowd, one surrenders one's autonomy in such a way that the illusion of *unlimited autonomy* can be projected.

In Kierkegaard's work, *The Concept of Dread* (see 1844/1975, pp. 101–105), we find that it is not merely the need to admit one's weakness and impotence that dissuades one from engaging one's individuality. To become an individual, one must come to know that which defies description, that which it is impossible to define or circumscribe in any way. Here one confronts the "possibility anterior to [particular] possibility" (p. 101), namely, the *freedom of spirit*. And this spirit inspires naught but dread: "The reality of the spirit …is a nothing which is able only to alarm" (p. 101).

In *The Concept of Dread*, Kierkegaard indicates certain stages in the process of individuation. For the animal and young child, spirit is largely dormant, peacefully asleep. "One does not…find dread in the beast, precisely for the reason that by nature the beast is not qualified by spirit" (p. 101). Then, with the loss of complete innocence and dawning of reflective consciousness, spirit progresses from a state of deep sleep to one of partial wakefulness. With spirit now "dreaming" (p. 101), the heady potentialities of its freedom are partially activated and an ambivalence takes hold. One is exhilarated by the unboundedness of the spirit, and, at the same time, repelled by it. The dread that arises with this glimpse of spirit's freedom is thus

> a *sympathetic antipathy and an antipathetic sympathy*.…One speaks of a sweet dread, a sweet feeling of apprehension, one speaks of a strange dread, a shrinking dread, etc.….If we observe [older] children, we find this dread more definitely indicated as a seeking after adventure, a thirst for the prodigious, the mysterious.…This dread belongs to the child so essentially that it cannot do without it; even though it alarms

> him, it captivates him nevertheless by its sweet feeling of apprehension. (1844/1975, p. 102)

Still later on, in entering adulthood, the person can experience "a heightened expression of dread, because this in a more profound sense is and is not, because in a more profound sense he loves it and flees from it" (p. 104). In its most developed form,

> One may liken dread to dizziness. He whose eye chances to look down into the yawning abyss becomes dizzy....[D]read is the dizziness of freedom which occurs when the spirit would posit the synthesis, and freedom then gazes down into its own possibility, grasping at finiteness to sustain itself. In this dizziness freedom succumbs....Dread is a womanish debility in which freedom swoons....But dread is at the same time the most egoistic thing, and no concrete expression of freedom is so egoistic as is the possibility of every concretion. This again is the overwhelming experience which determines the individual's ambiguous relation, both sympathetic and antipathetic. In dread there is the egoistic infinity of possibility, which does not tempt like a definite choice, but alarms and fascinates with its sweet anxiety. (1844/1975, p. 105)

In the foregoing passage, Kierkegaard is describing the climactic ambiguity that is reached when the process of individuation approaches completion. No longer merely dreaming, spirit now verges on complete wakefulness.

Until now, my account of Kierkegaard has neglected an important facet of his thinking—perhaps the *most* important aspect as far as he was concerned: his conviction as a Christian. The paradoxical truth to which one must leap by a passionate act of faith entails the "unity of the infinite and the finite" (Kierkegaard, 1846/1992, p. 38) that lies at the core of Christianity. To be sure, Kierkegaard was not speaking here of the kind of Christianity widely practiced by his contemporaries. In a series of pamphlets written near the end of his life, Kierkegaard launched a stinging attack against the insipid abstractions of organized Christianity, which he referred to as "Christendom." In Christendom,

> Everything is put in terms of reflection. The communication is qualified by reflection, hence it is indirect communication. The communicator is characterized by reflection, therefore he is negative—not one who says that he himself is a Christian in an extraordinary degree, or even lays claim to revelations (all of which answers to immediacy and direct communication); but, on the contrary, one who even affirms that he is

> not a Christian….The problem itself is a problem of reflection: To become a Christian…when one is a Christian of a sort. (1848/1975a, pp. 87–88)

Kierkegaard had little sympathy for the theologian or speculative philosopher who takes the truth of God to be a "what," an abstract object, and who seeks to approach that truth one small step at a time, through the painstaking pedantry of scholarly research. What does this "objective faith" mean?

> It means a sum of doctrinal propositions….The objective faith—it is as if Christianity also had been promulgated as a little system, if not quite so good as the Hegelian; it is as if Christ—aye, I speak without offense—it is as if Christ were a professor, and as if the Apostles had formed a little scientific society. Verily, if it was once difficult to become a Christian, now I believe it becomes increasingly difficult year by year, because it has now become so easy that the only ambition which stirs any competition is that of becoming a speculative philosopher. And yet the speculative philosopher is perhaps at the farthest possible remove from Christianity…. (1846/1975, pp. 119–120)

It appears that the *early* Christian was closer to genuine Christianity:

> In the earliest days the Christian was a fool in the eyes of the world….Now we are Christians as a matter of course, but if anyone desires to be a Christian with infinite passion he is judged to be a fool, just as it is always folly to put forth an infinite passionate exertion for the sake of becoming what one already is; as if a man were to sacrifice all his wealth to buy a jewel—which he already owned. Formerly a Christian was a fool in the eyes of the world, and now that all men are Christians he nevertheless becomes a fool—in the eyes of Christians. (1846/1975, p. 120)

Given his admiration for the fervor of original Christianity, was Kierkegaard espousing a simple return to those early days? Again, in Kierkegaard's own words, "'The Individual' is the category through which …this age, all history, the human race as a whole, must pass" (1859/1975, p. 100). The early Christian clearly had not yet passed through this category, was not yet fully individuated. In Part One, we found that the process of individuation is marked by a lessening of the tendency to project effective agency onto an external source, a transcendent power operating from beyond the human sphere. As the individual comes into his or her own, the gap between earth and heaven is thus closed. It is true that, with the advent of

Christianity, the finite and the infinite were brought closer together. Tarnas (1991) went so far as to say that, "In Christ…the historical and the timeless, the absolute and the personal, the human and the divine became one" (p. 102). However, the process of individuation was hardly complete in pre-Renaissance Christianity (see Chapter One). As a consequence, while the figure of Christ united the finite and the infinite in a symbolic fashion, it failed to do so *existentially*. Though Christ had walked among humans, he had not entered the heart of the individual person as Kierkegaard required, simply because, prior to the Renaissance, human individuality had not yet reached the necessary degree of development. Instead the person of Christ was externalized, projected onto a being who had indeed walked upon the earth, but who was actually more heavenly than earthly. This is not to say that the early Christian did not experience Jesus in a heartfelt manner. Listen to Kierkegaard's commentary on Christian and non-Christian emotion:

> The experience of being shaken, of being deeply moved, the coming into being of subjectivity in the inwardness of emotion, the pious pagan and the pious Jew have in common with the Christian. Upon this common basis of more universal emotion the qualitative difference must be erected and make itself felt, for the more universal emotion has reference only to something abstract: to be moved by something higher, something eternal, by an idea. And one does not become a Christian by being moved by something indefinitely higher, and not every outpouring of religious emotion is a Christian outpouring. (1846–1847/1975, p. 109)

The uniquely Christian emotion to which Kierkegaard alludes is emotion that can only be experienced by the person who has reached the stage of individuation where the projective externalization of divine agency can cease and the infinite can come to manifest itself concretely at the existential core of the finite. I suggest that such emotion could not have been experienced by the still weakly individuated *early* Christian, any more than by the "pious pagan" or "pious Jew." Therefore, although Kierkegaard certainly admired the passion of the early Christian when compared with the dispassionate smugness of contemporary Christendom, he could not have been calling for a simple return to early Christianity. It seems to me that—by grounding the infinite, by centering it squarely in the existential midst of the finite—Kierkegaard in fact was urging an unprecedented and radical *transformation* of Christianity.

Could we go so far as to say that—decades before Nietzsche, that other great early existentialist and champion of individuation—Kierkegaard was proclaiming the "death of God"? In his day, Nietzsche announced the "end of Christianity," maintaining that it "no longer has any sanction after it has tried to escape into some beyond" (1895/1975, p. 131). Is the "beyond" to which Nietzsche referred not closely akin to the religious abstraction that Kierkegaard had denounced earlier: "something higher, something eternal, …an idea" (1846–1847/1975, p. 109)? Perhaps the "death of God" then might be taken as the death of the abstracted, externalized deity that had prevailed for the many centuries that the process of individuation had been incomplete. When this process approached its culmination in the middle of the nineteenth century, Kierkegaard was able to recognize the limitations of abstract divinity and respond by insisting that deity be brought down to earth in earnest, that it come to life at the existential center of human being to complete human individuation in a bona fide way.

But we should be clear that, while Kierkegaard surely was advocating the transformation of Christianity, he was not calling for the transformation of God himself. Kierkegaard might have been persuaded to endorse Nietzsche's declaration that "God is dead," but he hardly would have gone as far with it. For Kierkegaard, the "death of God" could not have meant much more than the death of *humanity's misapprehension* of God. The status of God *per se* evidently was not negotiable. God was what He had always been in the rationalist tradition: utterly immutable, completely exempted from change. Did the presumed impermeability of God not maintain him after all as an *external* figure in relation to the all too mutable human being? Insusceptible to any sort of change, did God not remain an abstraction at the hands of Kierkegaard? Were God to descend fully into the human sphere so as to bring about a true "unity of the infinite and the finite" (Kierkegaard, 1846/1992, p. 38), would not God also be transmuted in the process? How could we speak of a bona fide merging of infinite and finite when the former continues to stand aloof?

Kierkegaard does tell us that the fusion of infinite and finite leads us beyond rationality and into absurdity or paradox. The paradox in question centers of course upon the person of Jesus Christ. God the Father may be immutable perfection, but His very own Son has a human side. Hence we have the paradoxical kinship of infinite and finite. However, the paradox loses much of its force when we realize that the bond between Father and Son is such that a certain distance is maintained between these relatives. Patriarchally conceived, the father-son relationship is one of blind submission: Asking no questions, the son must knuckle under to the unimpeachable authority of his father. When Kierkegaard declares that the "misfortune of our age…is disobedience, unwillingness to obey" (1846–1847/1975, p. 105), he adheres to the patriarchal model. This is especially clear in his commentary on the problem created when a son obeys his father in but a qualified manner:

> In case a son were to say, "I obey my father, not because he is my father, but because he is a genius or because his commands are always profound or clever"—then this filial obedience is affected. The son accentuates something which is entirely beside the point, he accentuates the cleverness and profundity in a *command*, whereas a command is precisely indifferent with regard to this qualification. The son is willing to obey by virtue of his father's profundity and cleverness, and by virtue of this it is precisely not possible to obey, for his critical attitude with regard to the decision whether the command is profound and clever undermines obedience. And this too is affectation when there is so much about accepting Christianity and believing in Christ on account of the profundity of the doctrine. One accepts orthodoxy by accentuating something which is entirely beside the point. (1846–1847/1975, p. 108)

The authoritarian father-son relationship that Kierkegaard is advocating here must apply *a fortiori* to the heavenly Father, and His Son. It is a strictly vertical relationship. God stands erect in sheer transcendence issuing his commandments from on high; His Son lies below, prostrating Himself at His Father's feet. This study in contrasts belies the notion that the infinite and the finite have been truly united in paradox. For while Jesus may indeed be His Father's Son, the asymmetric character of that authoritarian affiliation precludes the intimacy necessary for an authentic paradoxical union. In this regard, note Kierkegaard's comment that the "categories…of *immanence* and…of *transcendence*…belong each of them to their own qualitative spheres" (1846–1847/1975, p. 106). Despite their filial connection—or perhaps *because* of it—the Son is essentially immanent, whereas the Father is purely transcendent. But what of the *ordinary man*? Is *Christ* not transcendent in the eyes of the mere mortal? Yes, that would have to be the case, since man is the *son of Jesus* in this patriarchal lineage. Therefore, in the hierarchy that is conceived, while Jesus transcends man, He is immanent with respect to His own transcendent Father. Though Jesus does play this dual role here, immanence and transcendence still "belong each…to their own qualitative spheres." The finite and the infinite never really converge to bring about a full-fledged paradox.

Although Kierkegaard pulled no punches in his assault on the rationalist approach to God, what we are seeing is that his own God was at bottom that very same Deity, that same transcendent Subject. Evidently, it was not the anonymity of God that Kierkegaard denounced in his attack on the "crowd," but *human* anonymity. Man should make himself known to God, rather than having the impudence to strike a God-like pose of anonymous authority. Man should submit himself to his Father, offer himself up as object to this Divine Subject.

However, have we not seen that Kierkegaard emphasized the *subjectivity* of the human being? Was Kierkegaard not an *existentialist*? Was the knowledge of God not to be grounded in an unprecedented way by centering it within the existential midst of the finite individual? Did Kierkegaard not want the phrase "That Individual" to appear on his own tombstone? As far as I can tell, the basic problem is that Kierkegaard's writings are inconsistent on this score. An irreconcilable conflict appears in his work. On the one hand, there are his innovations as an existentialist, his groundbreaking challenge to the classical formula of object-in-space-before-subject. On the other hand, we have in Kierkegaard a Christian zealot whose authoritarian orthodoxy eliminates the paradox and reinstates the rule of the Divine Subject over His mortal objects. In this regard, the philosopher Walter Kaufmann observes that "Kierkegaard's thesis raises many questions" (1975, p. 85). Was Kierkegaard really speaking of the concrete subjectivity of human existence, wonders Kaufmann, "Or does [he] assume the objective truth of certain beliefs?" (1975, pp. 84–85)

2. "God is Dead"

Kierkegaard's successor, Friedrich Nietzsche (1844–1900), did go a great deal further in contributing to the "death of God." Like his predecessor, Nietzsche was passionately concerned with the process of individuation. What Nietzsche realized is that this process could never fully be brought to fruition without the utter demise of the system of values and beliefs that had stood in its way, namely, Christianity. It is not merely a matter of radically changing one's understanding of God, of taking God into one's heart, of accepting Divine Being at the existential core of mortal being. God Himself must radically change: He must be thoroughly undone. Nietzsche tells us that, in fact, at the time of his writing, the deed already had been accomplished, though, in another sense, it remained unconsummated because it had not yet been recognized. In one of the most renowned passages of nineteenth century literature, we learn from Nietzsche's "madman" the disruptive and exhilarating effects of this deicidal act:

> "Whither is God?" [the madman] cried; "I will tell you. *We have killed him*—you and I. All of us are his murderers. But how did we do this? How could we drink up the sea? Who gave us the sponge to wipe away the entire horizon? What were we doing when we unchained this earth from its sun? Whither is it moving now? Whither are we moving? Away from all suns? Are we not plunging continually? Backward,

> sideward, forward, in all directions? Is there still any up or down? Are we not straying as through an infinite nothing? Do we not feel the breath of empty space? Has it not become colder? Is not night continually closing in on us?....God is dead. God remains dead. And we have killed him....Is not the greatness of this deed too great for us? Must we ourselves not become gods simply to appear worthy of it? There has never been a greater deed; and whoever is born after us—for the sake of this deed he will belong to a higher history than all history hitherto."

And yet,

> "This tremendous event is still on its way, still wandering; it has not yet reached the ears of men....[D]eeds, though done, still require time to be seen and heard. This deed is still more distant from them than the most distant stars—*and yet they have done it themselves*." (1882/1974, Section 125)

Why is this deed so great? Again, it is essentially because the Christian God, more than any other force in contemporary culture, blocks the way to individuation. But the dangers accompanying God's demise clearly were not lost on Nietzsche. Indeed, in the end, he himself succumbed to them, both in his philosophy and in his personal life.

Before exploring the relationship between the "death of God" and Nietzsche's own eventual undoing, let us take note of an implication of the madman's litany that has special significance for us. He is portraying the bewilderment that results from "unchaining the earth from its sun." With the "entire horizon wiped away," we have lost our bearings and no longer know where we are going. "Are we not plunging continually? Backward, sideward, forward, in all directions? Is there still any up or down? Are we not straying as through an infinite nothing?" This total loss of orientation that attends the death of God betokens the demise of classical space. Of course, in the classical formula, space is inseparably linked to *subject* and *object*. So it is the whole traditional formulation of reality that is challenged by the death of God.

What we found in Part One is that the nineteenth century disruption of the old order exposed the long-repressed chaos of *apeiron*. The philosopher William Barrett (1958) confirms this resurfacing of the "Black Goddess" in his own assessment of existentialism. Barrett intimates that existentialist philosophy arose at a time when "certain unpleasant things the Enlightenment had dropped into the limbo of the unconscious have begun to backfire and have forced themselves finally upon the attention of modern man" (p. 276). What "unpleasant things"? Barrett speaks of "death, anxiety, guilt, fear and trembling, and despair" (p. 276), and he traces this irrational side of human nature back to "the Furies, the old goddesses of night and earth... [and] old Mother Earth herself" (p. 277). In Part One, we called this Black Goddess *apeiron*.

Now, if *apeiron* or the three Furies represent "the darker side of life" (Barrett, p. 278), life's "subterranean forces" (p. 279), the irrational impulses of the unconscious—nowhere in philosophy is *apeiron* better embodied than in the writings of Nietzsche. In the "dreams and visions that Nietzsche poured into his writings" (p. 184), he, "above all other philosophers, seemed to have found access to the unconscious" (p. 182). Barrett sees the Furies at play particularly in Nietzsche's *Thus Spoke Zarathustra*:

> [I]t is self-revelation at a greater, more primordial depth, where the stream of the unconscious itself gushes forth from the rock. Perhaps no other book contains such a steady procession of images, symbols, and visions straight out of the unconscious....[I]ts content is actually richer than Nietzsche's own conceptual thought, and its symbols of greater wisdom and significance than he himself was able to grasp. (Barrett, 1958, p. 187)

Nevertheless, while *apeiron* clearly made her presence felt in Nietzsche's work, it is equally clear that he was not content merely to immerse himself in the dark images that swirled about him in the aftermath of God's death. On the contrary, rather than dwelling in the shadows of the earth goddess, Nietzsche wanted to raise himself up to the status of a new god: "Must we ourselves not become gods?" the madman asks rhetorically. The task that Nietzsche thus set for himself was that of "rebirth and resurrection" (Barrett, 1958, p. 189). Given the nineteenth century collapse of classical culture, "how is man to be reborn, like the phoenix, from his own ashes? How is he to become really healthy and whole?" (Barrett, p. 189). That is to ask, how is the human being to restore himself and complete his project of individuation?

The question of human redemption actually is addressed in Nietzsche's work well before we hear it from the madman. Ten years prior to the publication of *The Gay Science* (1882/1974), Nietzsche's first book appeared. In *The Birth of Tragedy* (1872/1967), he recounts the pre-Socratic spirit of overflowing passion embodied in the wine god Dionysus. *The Birth of Tragedy* strikes a note of regret over the extinction of Dionysian exuberance by the cold intellect of Apollo—the god who in fact was a prototype of the *Christian* God. We can find in Nietzsche's earliest book the implication that redemption can come from *acknowledging* the irrational side of nature, from achieving a conscious recognition of *apeiron* that would bring her chaotic fury into balance with human reason. However, as Nietzsche's work evolved, "the Dionysian no longer signifies the flood of passion, but passion *controlled* as opposed to passion extirpated, the latter being associated with Christianity" (Kaufmann, 1968, p. 10; emphasis added).

This critical change in Nietzsche's thinking can be expressed in terms of a concept that was crucial for him, that of *causality*. There is no directed agency operative in *apeiron*; in itself, her frenzied outpouring of energy is "aimless," essentially acausal. Whereas the early Nietzsche paid homage to the acausal play of *apeiron* in the guise of the Dionysian, the later Nietzsche was more inclined to think of Dionysus as a principle of *self*-causation in which order is wrested from chaos.

To be sure, when Nietzsche proclaimed the death of God in his later work, he was announcing the undoing of *supernatural* agency. According to Nietzsche, Christianity is replete with "imaginary *causes*...[and] imaginary *effects*" (Nietzsche, 1895/1968, p. 581). For example, when a Christian experiences "pleasure in himself," when "he loves himself...he feels it—but precisely this love, this new self-valuation seems to him incredible; he can see in it only the wholly undeserved flowing down of a radiance of mercy from on high....[H]e conceives his mood of consolation as the effect upon him of an external power, the love with which fundamentally he loves himself appears as divine love" (1880/1992, p. 131). In recognizing the death of God, one can leave behind this "false psychology," this "aberration of reason and imagination" (p. 131), and come to realize that what had seemed like divine causation is in fact self-causation.

Nietzsche did not only deny the reality of supernatural agency, but rejected *all* concepts of external causation. Listen to what he has to say in his *Genealogy of Morals* about the notion of causality in general:

> A quantum of force is equivalent to a quantum of drive, will, effect—more, it is nothing other than precisely this very driving, willing, effecting, and only owing to the seduction of language...which conceives and misconceives all effects as conditioned by something that causes effects, by a "subject," can it appear otherwise. For just as the popular mind separates the lightning from its flash and takes the latter for an *action*, for the operation of a subject called lightning, so popular morality also separates strength from expressions of strength, as if there were a neutral substratum behind the strong man, which was *free* to express strength or not to do so. But there is no such substratum; there is no "being" behind doing, effecting, becoming; the "doer" is merely a fiction added to the deed—the deed is everything. (Nietzsche, 1887/1992, p. 118)

At first glance, Nietzsche might seem to be speaking about effects *without* causes. Of course, an event to which no cause of any kind can be attributed is a matter of happenstance, a chance occurrence beyond the ken of rational explanation, one transpiring in the province of *apeiron*. Is this what Nietzsche means? When he says that "the 'doer' is merely a fiction," that "the deed is everything," Nietzsche surely was attacking the substantialist notion of mechanical causation. The deeds in question—the "driving, willing, effecting"—clearly are not caused by external agents. But are these deeds simply without intelligible cause? Whereas the early Nietzsche might have been inclined toward an answer that would give more weight to *apeiron* as such, the Nietzsche who wrote *The Genealogy of Morals* evidently was not so much speaking of acausal deeds, as of deeds that are the causes of *themselves*, that shape themselves from the shapelessness of *apeiron*. Thus, for both the early and late Nietzsche we have *apeiron*, but his emphasis shifts in such a way that the dark goddess becomes *masculinized*, so to speak. She is turned into a force bent on taming the wildness in herself, constraining it, giving it sharp definition. We see this very clearly in *The Gay Science*, when Nietzsche proclaims:

> *One thing is needful*: "Giving style" to one's character….It is exercised by those who see all the strengths and weaknesses of their own natures and then comprehend them in an artistic plan until everything appears as art and reason and even weakness delights the eye. Here a large mass of second nature has been added; there a piece of original nature has been removed….Here the ugly which could not be removed is hidden; there it has been reinterpreted and made sublime….It will be the strong and domineering natures who enjoy their finest gaiety in such compulsion, in such constraint and perfection under a law of their own; the passion of their tremendous will relents when confronted with stylized, conquered, and serving nature; even when they have to build palaces and lay out gardens, they demur at giving nature a free hand. (1882/1974, section 290)

Is Nietzsche speaking of the human being's mastery over nature in this passage? For Nietzsche, the human being is an *integral part* of nature: "The fatality of his essence is not to be disentangled from the fatality of all that has been and will be….One is necessary, one is a piece of fatefulness, one belongs to the whole, one is in the whole" (Nietzsche, 1889/1968, p. 500). Therefore, the process of self-causation as exemplified in human activity can be naught but an instance of nature's own self-causation. To state the point succinctly and in quintessentially Nietzschean terms, nature—including human nature—is the *will to power*.

Nietzsche's late work, *The Will to Power* (1895/1992), provides what is perhaps the most vivid description of his "Dionysian world":

> This world: a monster of energy, without beginning, without end; a firm, iron magnitude of force that does not grow bigger or smaller, that does not expend itself but only transforms itself;...enclosed by "nothingness" as by a boundary; not something blurry or wasted, not something endlessly extended, but set in a definite space as a definite force, and not a space that might be "empty" here or there, but rather as force throughout, as a play of forces and waves of forces, at the same time one and many, increasing here and at the same time decreasing there; a sea of forces flowing and rushing together, eternally changing, eternally flooding back, with tremendous years of recurrence, with an ebb and a flood of its forms; out of the simplest forms striving toward the most complex, out of the stillest, most rigid, coldest forms toward the hottest, most turbulent, most self-contradictory, and then again returning home to the simple out of this abundance, out of the play of contradictions back to the joy of accord, still affirming itself in this uniformity of its courses and its years, blessing itself as that which must return eternally, as a becoming that knows no satiety, no disgust, no weariness: this, my Dionysian world of the eternally self-creating, the eternally self-destroying, this mystery world of the twofold voluptuous delight, my "beyond good and evil," without goal, unless the joy of the circle is itself a goal; without will, unless a ring feels good will toward itself—do you want a *name* for this world?....*This world is the will to power—and nothing besides!* (1895/1992, p. 106)

This "monster of energy" with "force throughout"; this "play of forces and waves of forces" that is "at the same time one and many"; this "sea of forces flowing and rushing together" with its "play of contradictions"; this "eternally self-creating...eternally self-destroying...mystery world...without goal ...without will"—all of this surely betokens the workings of *apeiron*. In *apeiron's* cycle of eternal recurrence, we have death as well as life, self-destruction along with self-creation; not only power but also its dissipation, not only will but its absence. Yet Nietzsche wants to call his Dionysian world the *will to power*. Evidently, it is because he wishes to stand for the affirmative in spite of *apeiron's* own impulse to *blend* affirmation and negation. "Life...strives after a *maximal feeling of power*" (1895/1992, p.106), and Nietzsche wishes above all else to affirm life. In another of his works, we can see how he agonized over the eternal cycle.

In *The Gay Science*, Nietzsche speaks of "The Greatest Weight":

> What, if some day or night a demon were to steal after you…and say to you: "This life as you now live it and have lived it, you will have to live once more and innumerable times more; and there will be nothing new in it, but every pain and every joy…will have to return to you….The eternal hourglass of existence is turned upside down again and again, and you with it, speck of dust!"
>
> Would you not throw yourself down and gnash your teeth and curse the demon who spoke thus? Or have you once experienced a tremendous moment when you would have answered him: "You are a god and never have I heard anything more divine." If this thought gained possession of you, it would change you as you are or perhaps crush you. The question in each and every thing, "Do you desire this once more and innumerable times more?" would lie upon your actions as the greatest weight. Or how well disposed would you have to become to yourself and to life *to crave nothing more fervently* than this ultimate eternal confirmation and seal? (1882/1992, p. 142–143)

Is Nietzsche urging unconditional acceptance of the eternal cycle in its totality and as such? Is the pain in life to be embraced for itself as readily as life's joy—the ultimate pain being that associated with death? Apparently not. Instead we are to endure the anguish of endless negation so that we may know eternally life's affirmation. It is therefore not a question of becoming well disposed toward negation and death in themselves. Rather, we are to be so "well disposed…to *life*" that we are willing to suffer through it again and again and suffer its termination, simply in order that we may *live*. For her part, *apeiron* does not value life over death. It is we who must make this choice. It is indeed a matter of value. Life must be valued so greatly that we are more than willing to die a thousand deaths that we may live that many lives. And valuing life means valuing *oneself*, Nietzsche tells us, which in turn means taking command of one's life instead of falsely attributing it to the action of external forces, be they mortal or divine.

Nietzsche believed that, by rejecting external causes and values in favor of self-valuation, he was overcoming the old metaphysical tradition. This is disputed by Heidegger (1943/1977). What is it that prevents Nietzsche from surpassing the classical tradition? It is precisely his incessant "value-positing" (Heidegger, 1943/1977, Part Two).

Heidegger observes Nietzsche's deep preoccupation with the question of values: "the establishing of values, the devaluing of values, the revaluing of values…the positing of values anew and…the principle of all value-positing—a positing that values differently. The highest purposes, the grounds and principles of whatever is, ideals and the suprasensory, God and the gods—all this is conceived in advance as value" (1943/1977, p. 70).

Therefore, says Heidegger, "A sufficiently clear exposition of what Nietzsche thinks in the word 'value' is the key to an understanding of his metaphysics" (p. 70).

What then does Nietzsche mean by "value"? "Nietzsche says in a note (1887–88) what he understands by value: 'The point-of-view of "value" is the point-of-view constituting the *preservation-enhancement conditions* with respect to complex forms of relative duration of life within becoming' (*Will to Power*, Aph. 715)" (Heidegger, 1943/1977, p. 71). Implicit in Nietzsche's reference to the "point-of-view of 'value'" is that,

> The essence of value lies in its being a point-of-view. Value means that upon which the eye is fixed. Value means that which is in view for a seeing....This seeing is of such a kind that it sees inasmuch as...it has set before itself and thus posited what is sighted, as a particular something. (Heidegger, 1943/1977, pp. 71–72)

That is to say, a value, in its essence, is an *ob-ject*; it is that which is cast before the eyes of a seer, who is the *subject*. This is true even though, with Nietzsche, the subject itself becomes an object—that is, the subject places the highest value on *itself*. Thus, valuing is objectifying, and, as Heidegger puts it,

> Objectifying, in representing, in setting before, delivers up the object to the *ego cogito*. In that delivering up, the *ego* proves to be that which underlies its own activity (the delivering up that sets before), i.e., proves to be the *subiectum*. The subject is subject for itself. The essence of consciousness is self-consciousness. Everything that is, is therefore either the object of the subject or the subject of the subject. (1943/1977, p. 100)

Therefore, with Nietzsche,

> Man...rises up into the subjectivity of his essence. Man enters into insurrection. The world changes into object. In this revolutionary objectifying of everything that is, the earth, that which first of all must be put at the disposal of representing and setting forth, moves into the midst of human [value] positing and analyzing. The earth itself can show itself only as the object of assault....Nature appears everywhere ...as the object of technology. (Heidegger, 1943/1977, p. 100)

Or we can say that Nature—in the form in which it has been "masculinized," subjectivized, or anthropomorphized by Nietzsche—becomes an object to *itself*. It is clear from this why Nietzsche fails to overcome the metaphysical tradition. For said tradition is governed by the old formula, object-in-space-before-subject—a formula to which Nietzsche expressly adheres and, in fact, brings to fruition. At the hands of Nietzsche, the "object," that is, the *value*, now lies in the subject per se, and this carries subjectivity into its most essential expression, since "the essence of consciousness is self-consciousness" (Heidegger, 1943/1977, p. 100).

Where does the will to power fit in here? Self-valuation is the *modus operandi* of the will. Heidegger elucidates this by focusing on Nietzsche's definition of value. To repeat, "'The point-of-view of "value" is the point-of-view constituting the *preservation-enhancement conditions* with respect to complex forms of relative duration of life within becoming' (*Will to Power*, Aph. 715)" (Heidegger, 1943/1977, p. 71). By the "preservation-enhancement" of life, Nietzsche means that life functions in such a way that it does not merely preserve itself but also, in so preserving, it necessarily enhances itself. And the basic condition to be met for achieving this preservation-enhancement is the positing of values (objectification). For Nietzsche then, "life, in its essence, proves to be value-positing" (Heidegger, 1943/1977, p. 73). We already have seen that Nietzsche views life as synonymous with the will to power. Heidegger confirms this in his further comment on Nietzsche's definition: "When Nietzsche concludes his characterization of the essence of value with the word 'becoming,' then this closing word gives the clue to the fundamental realm within which alone values and value-positing properly belong. 'Becoming' is, for Nietzsche, the 'will to power.' The 'will to power' is thus the fundamental characteristic of 'life'" (1943/1977, p. 74).

However, when Heidegger claims that, "the modern metaphysics of subjectness is consummated in Nietzsche's doctrine of the will to power" (1943/1977, p. 83), this does not necessarily mean that the doctrine was actually brought to fruition by Nietzsche himself. To carry metaphysics to its consummation evidently would require an "affirmation of life" effective enough to deny death more completely than Nietzsche in fact was able to do. Nietzsche's limitation in this regard is nicely brought out in Barrett's interpretation of *Thus Spoke Zarathustra* (1958, pp. 187–197). The section of this poetic work that Barrett focuses upon is titled, "On the Vision and the Riddle" (Nietzsche, 1878/1968, pp. 267–272).

Barrett first tells us that it is through Zarathustra that Nietzsche attempts to work out the problem of individuation: of "how man is to be reborn," and of how, in this resurrection, "the superior individual…is to be formed" (1958, p. 189). Zarathustra is to overcome the "Spirit of Gravity" (p. 192); he is to scale the heights of perfection by ridding himself of all human weakness and illusion. Zarathustra is the personification of Nietzsche's will to power, the Faustian prototype of his new individual—the "overman" or "superman" who is to break the shackles of ignorance that had been forged by Christianity and metaphysics. But, as Zarathustra's vision of ascent unfolds, he encounters admonitions and dark omens. First there is the "dwarf" who sits on his back as he climbs his mountain. This homunculus weighs him down, and warns him mockingly:

> O Zarathustra…you philosopher's stone! You threw yourself up high, but every stone that is thrown must fall. O Zarathustra, you philosopher's stone, you slingstone, you star-crusher! You threw yourself up so high; but every stone that is thrown must fall. Sentenced to yourself and to your own stoning—O Zarathustra, far indeed have you thrown the stone, but it will fall back on yourself. (Nietzsche, 1878/1968, p. 268)

Yet Zarathustra defies the dwarf's warning: "Dwarf! It is you or I!….But I am the stronger of us two" (p. 269).

In the next part of Zarathustra's visionary dream, he and his dwarf happen upon a gateway at which two paths meet: "This long lane stretches back for an eternity. And the long lane out there, that is another eternity" (p. 269). Elaborating upon these words, Zarathustra intimates the idea of eternal recurrence, after which he expresses his misgivings: "Thus I spoke, more and more softly; for I was afraid of my own thoughts and the thoughts behind my thoughts" (p. 270).

Finally Zarathustra encounters the most chilling omen of all. A dog howls in fear under a full moon to announce the appearance of a man who is lying on the ground: "A young shepherd I saw, writhing, gagging, in spasms, his face distorted, and a heavy black snake hung out of his mouth. Had I ever seen so much nausea and pale dread on one face? He seemed to have been asleep when the snake crawled into his throat, and there bit itself fast" (p. 271). Feeling dread and nausea himself, Zarathustra cries out to the shepherd: "Bite! Bite its head off! Bite!" (p. 271). And the shepherd "bit as my cry counseled him….Far away he spewed the head of the snake—and he jumped up. No longer shepherd, no longer human—one changed, radiant, *laughing*! Never yet on earth has a human being laughed as he laughed! O my brothers, I heard a laughter that was no human laughter" (Nietzsche, 1878/1968, p. 272).

What does Barrett make of all this? Near the end of his account, Zarathustra poses a riddle to his readers: "*Who* is the shepherd into whose throat the snake crawled thus? *Who* is the man into whose throat all that is heaviest and blackest will crawl thus?" (Nietzsche, 1878/1968, p. 272). Barrett takes up Zarathustra's challenge by proposing that the man in question is none other than Nietzsche himself:

> [B]oth the serpent and the dwarf set for [Nietzsche] the same task: to acknowledge "the heaviest and the blackest in himself." We commonly speak of the truth as a bitter pill that we have to swallow, but the truth about ourselves may take even the more repulsive form of a reptile. Nietzsche does not swallow the serpent's head; he denies his own shadow, and out of it he sees a transfigured being spring up. This being laughs with a laughter that is no longer human. We know this laughter all too well: it is the laughter of insanity....
>
> There is an inner coherence in the vision of Zarathustra, in that each of its three parts—the dwarf, the Eternal Return, and the shepherd spitting out the serpent—presents an obstacle and objection to Nietzsche's utopian conception of the Superman. They prefigure his own personal catastrophe [his insanity]; but since he was a thinker who really lived his thought, they indicate the fatal flaw in all such utopian thought. He who would launch the Superman into interstellar space had better recognize that the dwarf goes with him. (1958, pp. 195–196)

Now, we considered above Barrett's introduction to *Thus Spoke Zarathustra*. Here he acknowledges that "[n]o adequate psychological commentary...has yet been written" (1958, p. 187) on this book. "It is a unique work of self-revelation but not at all on the personal or autobiographical level...it is self-revelation at a greater, more primordial depth, where the stream of the unconscious itself gushes forth from the rock....[I]ts content is actually richer than Nietzsche's own conceptual thought, and its symbols of greater wisdom and significance than he himself was able to grasp" (p. 187). In these comments, Barrett appears to be adopting a perspective akin to that of Jung (though Jung is not cited). He seems to be suggesting that the unconscious realm that is tapped in *Zarathustra* is *transpersonal*, and, therefore, that the meaning of its imagery goes beyond the particular concerns of Friedrich Nietzsche the historical personage, that it touches on themes of universal import for humanity. If the transpersonal dimension does operate in *Zarathustra*, might it not be especially evident in the highly potent symbols found in the section of this book we have been focusing upon? In his treatment of the "Vision and the Riddle," Barrett's only direct reference to the archetypal domain comes right after he asks why the circle of "Eternal Return had such a powerful hold on Nietzsche's emotions." At this point he notes that "[t]he circle is a pure archetypal form for the eternal" (p. 194). Would a closer examination of Zarathustra's vision reveal other transpersonal features?

In point of fact, the material appearing in the "Vision and the Riddle" is replete with highly charged alchemical symbols and imagery. This is evident from Jung's own explicit analysis of Nietzsche's *Zarathustra*. Before turning to the evidence, let us note Jung's general fascination with Nietzsche, whom he regarded as an important figure in the hermetic tradition. What we learn from Jung is that alchemy is not merely a practice that aims at transmuting base metals into precious ones by magical means; rather, its essential aim is the transmutation of the *human being*, both on a personal and transpersonal level. In alchemy, "the superior individual…is to be formed," to echo Barrett's characterization of the basic goal that *Nietzsche* set for himself. Alchemy is thus a quest for the "superman," for the Self, for individuation fulfilled—and, according to Jung, Nietzsche played a significant role in this venture. In his *Alchemical Studies* (1967), Jung traces a direct line of descent from the sixteenth century alchemist Paracelsus to Goethe to Nietzsche: "Not for nothing was Paracelsus the prototype of [Goethe's] Faust….From Faust the line leads directly to Nietzsche, who was a Faustian man if ever there was one" (p. 118). And, "It was from the spirit of alchemy that Goethe wrought the figure of the 'superman' Faust, and this superman led Nietzsche's Zarathustra to declare that God was dead and to proclaim the will to give birth to the superman, to 'create a god for yourself…'" (p. 128).

Archetypal images and symbols function critically in the process of alchemical transformation. This is the main theme of Jung's *Symbols of Transformation* (1956), and it is here that he offers an analysis of the shepherd and snake found in *Zarathustra's* "Vision and Riddle":

> [T]he snake represents the unconscious psyche which, like the snake-god in the Sabazios mysteries, crawls into the mouth of the celebrant, i.e., Nietzsche himself as the…shepherd of souls and preacher, firstly to stop him from talking too much, and secondly to make him… "enthused," "filled with God."….The unconscious insinuates itself in the form of a snake if the conscious mind is afraid of the compensating tendency of the unconscious, as is generally the case in regression. But if the compensation is accepted in principle, there is no regression, and the unconscious can be met half-way…. (p. 379)

However, in the case of Nietzsche's shepherd, "the compensation appears in a form that cannot be accepted," so the shepherd spits the snake out instead of incorporating it. "Cases of this kind occur when the unconscious has been resisted for too long on principle, and a wedge violently driven between instinct and the conscious mind" (Jung, 1956, p. 380). Does this mean that Nietzsche was relatively intransigent when it came to "meeting the unconscious half-way"? On the contrary, Nietzsche, in serving as the premier *alchemist* of his time, confronted the unconscious far more directly and persistently than did the vast majority of his contemporaries. It therefore seems more appropriate to attribute his inability to "meet the unconscious halfway" to the overwhelming power of the archetypal images he encountered than to his own particular intransigence or weakness as a person. Yes, Nietzsche did succumb; in the end he went mad. But were his "sane" contemporaries—who resisted the "serpent" to the point of never even consciously confronting it—really "healthier" than he? In speaking of cases "when the unconscious has been resisted for too long," Jung apparently speaks as a psychoanalyst, and, in this capacity, he refers to the personal histories of certain patients. From the broader perspective of Part One, we saw in effect that the "unconscious" can be resisted by an *entire culture*, in which case the resistance indeed can be protracted and exceedingly difficult to overcome. I submit that Nietzsche's failure to "meet the unconscious halfway" reflects the failure of nineteenth century culture to do this. I suggest that the snake that haunted Zarathustra was the self-same reptilian creature that plagued science, art, and the media in that historical period. For, the serpent is a primal embodiment of *apeiron*. To appreciate more concretely Nietzsche's archetypal battle with the Black Goddess, let us consider the imagery of Zarathustra's vision in greater detail.

In recounting his experience, Zarathustra tells us of how he "ascended defiantly through stones" (Nietzsche, 1878/1968, p. 268):

> Striding silently over the mocking clatter of pebbles, crushing the rock that made it slip, my foot forced its way upward. Upward—defying the spirit that drew it downward toward the abyss, the spirit of gravity, my devil and archenemy. Upward—although he sat on me, half dwarf, half mole, lame, making lame, dripping lead into my ear, leaden thoughts into my brain. (p. 268)

It is at this point that Zarathustra's homunculus taunts him: "O Zarathustra …you philosopher's stone! You threw yourself up high, but every stone that is thrown must fall….O Zarathustra, far indeed have you thrown the stone, but it will fall back on yourself" (p. 268). Zarathustra is striving for light. He seeks the brightness of the sky above the darkness of stone. He wants to lighten his spirit by transcending the pull of gravity. But his homunculus reminds him that, "what goes up must come down." Though Zarathustra wishes to climb the highest mountain, to thrust himself straight into the heavens, his lame dwarf warns him of the one-sidedness of this quest: "'All that is straight lies,'" says the homunculus. "'All truth is crooked; time itself is a circle'" (p. 270). Here the dwarf is intimating the cycles of nature in which dark and light, down and up, earth and sky, death and life, are inseparably bound together. The fundamental message is that Zarathustra's desire to break the circle open—to stretch it into a line wherein death and life are poles apart—is basically misguided.

When the dwarf names Zarathustra as the "philosopher's stone," the alchemical nature of the passage is explicitly disclosed. For the whole alchemical enterprise revolves around this mysterious substance. In alchemy, the *lapis philosophorum* is the lowest and the highest, the alpha and omega, the beginning of the opus and its end. Although the "arcane substance" is primitive and undeveloped—"prima materia," the "primeval chaos" (Jung, 1967, p. 235)—it is also "the panacea, the universal medicine…the tincture that transmutes base metals into gold and gravel into precious stones" (Jung, 1967, p. 101). Thus Jung notes that "The term *lapis* is used all through the [alchemical] literature for the beginning and the goal" (1968, p. 320n). Elsewhere Jung relates the work of alchemy to the philosopher's stone as follows:

> [T]he alchemical projections represent collective contents that stand in …compensatory relation…to our highest rational convictions and values. They give the strange answers of the natural psyche to the ultimate questions which reason has left untouched. Contrary to all progress and belief in a future that will deliver us from the sorrowful present, they point back to something primeval, to the apparently hopelessly static, eternal sway of matter….They show us, as the redemptive goal of our active, desirous life, a symbol of the inorganic—the stone. (1967, p. 238)

When Jung observes that "unconscious nature…produced the image of the lapis," he refers to nature as "she" (Jung, 1967, p. 95). This original chaos or *prima materia* evidently is *maternal* in character. Alchemy goes beyond the abstractions of Christianity to "add a fourth, feminine principle to the masculine Trinity. The result is a quaternity, which forms a real and not merely postulated symbol of totality" (1967, p. 96). Four-fold symbols of this kind often appear as "squared circles" (p. 96). In such alchemical structures, the "accentuation of matter [mater, matrix, mother] is above all evident in the choice of the *stone* as a God-image" (p. 97).

Why is the quaternary lapis "a real and not merely postulated symbol of totality"? Grounded in the reality of nature, the stone incorporates nature's concrete oppositions without simply dissolving them in favor of an idealized unity. For example, in the quaternary conjunction of earth, fire, water, air, the elements do come together to form an indivisible whole, and yet, at the same time, each element retains its own identity. So, instead of signifying a single abstract unity that is imposed from above, the stone connotes a *para-*

doxical unity-in-diversity. Related paradoxes are found throughout the alchemical literature (see Rosen, 1995) and apply to the most fundamental oppositions, such as that between good and evil, light and dark, and life and death. In this regard, compare Nietzsche's seemingly one-sided affirmation of life—symbolized by Zarathustra's monomaniacal quest for the light—with a statement on "the significance of the *lapis*...ascribed to Hermes ....'Understand, ye sons of the wise, what this exceeding precious stone proclaims....I beget the light, but the darkness too is of my nature'" (Jung, 1968, p. 79). Of course, the homunculus in the "Vision and the Riddle" informs us that *Zarathustra himself* is the philosopher's stone, though Zarathustra apparently cannot accept the message. We can see from this that Zarathustra/Nietzsche truly was at war with himself. For his part, Jung was well aware of the alchemical symbolism of the stone in Nietzsche's *Zarathustra*. In fact, we learn from Jung that Zarathustra's homunculus is itself also a fundamental aspect of the stone: "In [Nietzsche's] stone sleeps the spirit *Mercurius*, the 'circle of the moon,' the 'round and square,' the homunculus, Tom Thumb and Anthropos at once, whom the alchemists also symbolized as their famed *lapis philosophorum*" (1969, p. 304).

Jung places emphasis on the *circular* nature of the stone. The lapis is seen as a "dark sphere" associated with the "*massa confusa*" (Jung, 1968, p. 325), the original chaos of the world. Jung speaks of "the perfect, 'round' nature of the *lapis*, which arises from, and constitutes, the primal sphere" (p. 325). Later on, he relates the *prima materia* to the "wheel of creation," the "*opus circulatorium*" (p. 381) that defines the basic work of alchemy. Later still, Jung notes that the lapis is signified in the wheels of Ezekiel's vision (p. 385), and that, "The idea of the circulatory *opus*, or rotating arcane substance, finds expression as early as Komarios, who speaks of the 'mystery of the whirlwind in the manner of a wheel'" (p. 386).

Perhaps we have now seen enough to be able to recognize the philosopher's stone for what it is: It is the incarnation of Mother Nature, of the Black Goddess, of chaotic *apeiron*. It is the trans-rational "feminine Physis" (Jung, 1968, p. 304) who—like a black hole in space—draws into her cyclonic roundness all of nature's oppositions, embracing all, denying none. The spherical goddess reasserts herself in *each* stage of Zarathustra's vision, much to his consternation. She is the circle of eternal recurrence encountered at the gateway in the second part of the dream—a circle that Zarathustra fears and would love to unknot. She is the full moon at which the terrified dog howls in part three of the vision, and is the "heavy black snake" that hangs from the shepherd's mouth. A snake indeed! For it is clear from numerous references, allusions, and illustrations in Jung's volumes on the hermetic tradition (1967, 1968, 1970) that alchemy equates the philosopher's stone with Mercurius, the Mercurial Serpent or *uroboros*. Jung characterizes this snake that forms a circle by devouring itself, by swallowing its own tail (Fig. 5), as "the primal symbol of alchemical truth" (1970, p. 102n).

Figure 5. The Uroboros, from the *Chrysopoeia* of Cleopatra, Alexandrian Period of Ancient Egypt

In declaring that Zarathustra himself is the philosopher's stone, the dwarf is hinting that he is also the uroboros, the self-swallowing snake, she who destroys and recreates herself in a single gesture. This is what Zarathustra/Nietzsche finds so impossible to swallow, what nauseates him and fills him with fear. Impelled by the phallic thrust toward light and life that marks the patriarchal project of individuation, Nietzsche must deny at all costs his participation in the serpentine cycle of life and death. "Bite! Bite its head off!" cries Zarathustra. With this the uroboros is decapitated. The price for this violence toward nature is Nietzsche's sanity. In denying the "heavy black snake," in defying the gravitational pull of the black hole, the balance of nature is upset and the "spirit of gravity" is displaced by its diametrical opposite: an eerie levity, the mad laughter of the shepherd-become-superman. Of course, the very madness of this levity has a gravity of its own. The airiness of reason, in being brought to its most extreme idealization, takes on the heavy cast of dementia.

If Nietzsche went mad, and if his madness resulted from repressing the bodily rhythms of nature and becoming lost in the rarefied heights of self-absorbed abstraction (as well as from his syphilis!), then the madness of Nietzsche was mild compared with what was to follow. Nietzsche in fact may be regarded as a *transitional* figure. Having studied *Zarathustra*, we are now able to see why it was that—even though "the modern metaphysics of subjectness is consummated in Nietzsche's doctrine of the will to power"

(Heidegger, 1943/1977, p. 83)—Nietzsche himself could not fully realize this doctrine. It was because of the relentlessness of his "dwarf."

Again, the central problem for Nietzsche was that of "rebirth and resurrection" (Barrett, 1958, p. 189). After the collapse of classicism and resurfacing of *apeiron* at mid-nineteenth century, Nietzsche took upon himself the task of putting the Spirit Mercurius "back into the bottle." He thus faced the challenge of reenacting history. Once more it became necessary to wrestle with tempestuous *apeiron*, as Plato's Demiurge had done two thousand years before (see Chapter One). Once more the "mountain" had to be climbed, as Petrarch had done at the outset of the Renaissance (Chapter One, p. 10). But it is clear from Zarathustra's grimly Herculean efforts that a far "higher mountain" was now to be scaled, one that reached its peak in a subjectivity so "stratospheric" that the terrain below could barely even be seen. In fact, there was to be *only* this subject. He was to be his *own* object, to act solely upon himself, and for himself. So high was this new individual to climb that, eventually, he would free himself completely from the earth's atmosphere and enter a state of "weightlessness"; breaking the hold of gravitation, this superman would *levitate*. It would be in this way that "Man… [would rise] up into the subjectivity of his essence"—into pure self-consciousness—since "the essence of consciousness is self-consciousness" (Heidegger, 1943/1977, p. 100). Nietzsche, however, could not raise himself that high, thanks to the persistence of his "dwarf." The "modern metaphysics of subjectness" had thus to be consummated by others. —Which brings us back to Copenhagen.

At the beginning of this chapter, I noted Kierkegaard's positive influence on Bohr. However, despite the fact that Bohr found Kierkegaard's *Stages on Life's Way* to be "'one of the most delightful things I have ever read,'" he went on to declare that he did not agree "with the antiscientific attitude of much of the work" (Holton, 1988, p. 129). Indeed, Bohr—like his colleague Albert Einstein, was a member of the next generation. He was an important participant in the expedition to reach the "mountaintop," to establish a new order of scientific objectivity for the newly conceived subjectivity —the more abstract, self-active, self-conscious modernist subject who came to prominence in the wake of "God's death." Of course, unlike Kierkegaard and Nietzsche, Bohr was not primarily a philosopher. But just as physics was establishing its own version of modernist subjectivity as the twentieth century commenced; and just as art, the media, and other spheres of culture were following suit; the same appeared to be happening in the field of philosophy. With advances in the modernist philosophy of consciousness, the passion of existentialism was being reined in by a cooler, more "scientific" approach. A prime example of this is the philosophy of Edmund Husserl (1859–1938), founder of *phenomenology*.

3. "To the Things Themselves"

Husserl picks up where Nietzsche left off. His phenomenological philosophy aims at nothing less than a new consciousness, a purely reflective, self-active consciousness that ultimately would remake not only Western philosophy, but humanity itself. It is not just Husserl's vision of the future that is reminiscent of Nietzsche, but his reading of the present state of affairs in philosophy as well. From the beginning of his career to the end, Husserl tirelessly criticized what he viewed as the primary impediment to the advancement of philosophy: naturalistic objectivism, with its doctrine of mechanical causes.

In a 1911 essay, "Philosophy as Rigorous Science," Husserl begins his description of "naturalistic philosophy" in the following way:

> [T]he naturalist...sees only nature, and primarily physical nature. Whatever is, is either itself physical, belonging to the unified totality of physical nature, or it is in fact psychical, but then merely as a variable dependent on the physical, at best a secondary "parallel accompaniment." Whatever is belongs to psychophysical nature, which is to say that it is univocally determined by rigid laws. (1911/1965, p. 79)

What laws are these? They are the laws of external causation as posited by mechanistic science. Thus Husserl states that the naturalistic order projects appearances of "real things" that

> stand there as incorporated in the totality of one corporeal world that binds them all together, with its one space and its one time....[O]nly in the causal relation to or connection with each other do they retain their individual identity (substance), and this they retain as that which carries "real properties." All physically real properties are causal....Each physical thing has its nature (as the totality of what it, the identical, is) by virtue of being the union point of causalities within the one all-nature. Real properties (real after the manner of things, corporeal) are a title for the possibilities of transformation of something identical, possibilities preindicated according to the laws of causality. (1911/1965, p. 104)

Now, when Husserl notes that the "real things" projected by naturalism "stand there," he means that they are naïvely taken to be simply autonomous. "All natural science is naïve....The nature that it will investigate is for it simply there" (Husserl, 1911/1965, p. 85). And, "natural science accepts nature as given" (p. 87). In so doing, it "pretends at every step to posit and to know a nature that is in itself—in itself in opposition to the subjective flow of consciousness" (p. 88). As a consequence, "without realizing it, naturalism refutes itself" (p. 80). For,

> The naturalist is, one can safely say, idealist and objectivist in the way he acts. He is dominated by the purpose of making scientifically known ...whatever is genuine truth....He is, however, an idealist who sets up and (so he thinks) justifies theories, which deny precisely what he presupposes in his idealist way of acting. (1911/1965, pp. 80–81)

In other words, naturalism contradicts itself because it operates subjectively to project an external reality that is then assumed to be simply independent of its subjective action. "The same is true in regard to nature in the broader, psychophysical sense, or in regard to the sciences that investigate it—in particular, therefore, in regard to psychology. The psychical does not constitute a world for itself...[but] reveals itself empirically as bound to certain physical things called bodies" (p. 85).

Husserl sees only one remedy for the "absurdity of a theory of knowledge based on natural science" (p. 88):

> [T]he investigation must be directed toward a scientific essential knowledge of consciousness, toward that which consciousness itself "is" according to its essence in all its distinguishable forms....Every type of object that is to be the object of a rational proposition, of a prescientific and then of a scientific cognition, must manifest itself in knowledge, thus in consciousness itself....[O]bjectivity...must precisely become evident purely from consciousness itself. (1911/1965, pp. 89–90)

With these words Husserl adumbrates the science of "pure consciousness" that he calls *phenomenology*. Here,

> we meet a science of whose extraordinary extent our contemporaries have as yet no concept; a science, it is true, of consciousness that is still not psychology; a phenomenology of consciousness as opposed to a natural science about consciousness....[P]sychology is concerned with ...consciousness from the empirical point of view, as an empirical being in the ensemble of nature, whereas phenomenology is concerned with "pure" consciousness.... (1911/1965, p. 91)

What is the implication of asserting that while natural science is "about" consciousness, phenomenology is "of" consciousness? By presupposing consciousness to be contained in an outer domain of mechanical causation, natural science loses direct access to consciousness. Phenomenology would set aside the naturalistic presupposition and attend "*to the things themselves*," as the slogan of phenomenology goes; that is, phenomenology would adhere to the *phenomena* of experience, to just what is given to us in experience *as* experience, as a datum of consciousness, and not as something standing "out there" on its own, in a simply autonomous natural world. Husserl recognizes that we "do not easily overcome the inborn habit of living and thinking according to the naturalistic attitude, and thus of naturalistically adulterating the psychical" (p. 109). Diligent effort therefore is required if we are to learn to "take phenomena as they give themselves" (p. 108), thereby resisting the tendency to "attribute a nature to phenomena, [and to proceed] to investigate their...causal connections" (p. 106).

In denying causal efficacy to the phenomena, is Husserl rejecting active agency altogether? On the contrary, though the phenomena of consciousness may not be active in the sense of external influences, "The psychical is not simply experienced as something that appears; it is 'vital experience' and vital experience seen in reflection; it appears as itself through itself, in an absolute flow" (p. 107). Husserl likens the phenomena to "monads that have no windows and are in communication only through empathy" (p. 106). In a footnote to this passage, translator Quentin Lauer observes that, by adopting the Leibnizian notion of self-active, self-causative monads, Husserl is by no means endorsing Gottfried Leibniz's metaphysical system: "He simply borrows the image to emphasize the denial of any *external* causality in the realm of the psychic. Nothing was more distasteful to him than the mechanistic psychology of Locke" (Lauer, 1965, p. 106, n. 43; emphasis added). In a previous footnote, Lauer comments that while Husserl rejects the doctrine of external causes, he "does speak of a 'spiritual causality'" (1965, p. 104, n. 42). What Husserl is saying then, in his more sober scientific prose, is what Nietzsche said before him more poetically: that consciousness neither is subject to outside causation nor is it acausal, but is the cause of itself; that it involves neither exterior action nor passivity but *self*-action; that it entails the application of the will to itself, as in Nietzsche's will to power.

To the extent that the phenomenologist succeeds in overcoming the naturalistic attitude, he or she thus can participate in the self-activity of consciousness. Consciousness indeed does not merely observe itself, as it is purported to do in the method of introspection that Husserl rejects. Rather, in its intentionality, in all of the acts whereby it makes meaning for itself by establishing itself in relation to meaningful others (objects), consciousness *actively constitutes* itself. And the primary aim of phenomenological reflection is to directly take part in this self-constitution.

The grasping of consciousness by consciousness itself is to be achieved through "essential intuition": "[I]f phenomena have no nature [are not naturalistic], they still have an essence, which can be grasped and adequately determined in an immediate seeing...an essential intuition" (Husserl, 1911/1965, p. 110). According to Husserl, "just as immediately as one can hear a sound, so one can intuit an 'essence'" (p. 115). However, this "depends on ...not confounding phenomenological intuition with 'introspection,' with interior experience—in short, with acts that posit not essences but individual details corresponding to them" (p. 115). It is because introspective psychology engages in a mere "existential positing of [psychical] nature," in an "investigation of being-there" (p. 116), that its sight is limited to the "individual details" of "interior experience." When the psychic world is naturalized in this way, it can be manifested only in the form of finite particular beings, objects-in-space. But phenomenology cannot rest content with the particularities of existence, given their inherent variability, their ambiguous and irrational character. For—as the title of Husserl's essay makes clear—phenomenology is to be a thoroughly "rigorous science." As such, it must aspire to absolute validity, something it can only achieve by forgoing particular existence in favor of universal essence. Hence, "For phenomenology, the singular is always *apeiron* [!]. Phenomenology can recognize with objective validity only essences and essential relations" (p. 116). Commenting accordingly on Husserl's privileging of essentialism over existentialism, Lauer says of the former that it "has no concern for the individual, the unique, precisely because it cannot do so and still be 'scientific'" (p. 117, n. 57). Given its aim of pure scientific rationality, phenomenology must turn its intuitive eye toward that which can be known with absolute certainty. For Husserl, "whatever makes for uncertainty must be eliminated from philosophical consideration" (Lauer, 1965, p. 64). (Notice the striking contrast between Husserl's cool dismissal of *apeiron* and Nietzsche's lifelong struggle with it.)

There is only one thing that consciousness can come to know with absolute certainty: *itself*. This assumption lies at the very heart of Husserl's new discipline and is implied by its name. Phenomenology limits the study of consciousness to the *phenomena*, to that which is *immanent* to consciousness and thus can be given to it in its immediacy—as opposed to what is *noumenal*, to the extra-mental reality that classical metaphysics alleges to lie "beyond mere appearances." It is the phenomena, the "things themselves," that we are to grasp through essential intuition. By what specific methodology is this to be achieved?

Lauer (1965, p. 27) summarizes the procedures originally set forth by Husserl in *Logical Investigations* (1900). In part, the phenomenological method must be negative: it must eliminate what stands in the way of "immanent seeing" (Husserl, 1911/1965, p. 108). The old metaphysical tendency to locate meaning in the "outside world of nature" surely is powerful. Heroic efforts therefore are required in order to break the hold of the natural attitude. Through the procedure of *epochē*, one attempts to bracket or suspend the habitual commitment to external reference whenever and wherever it arises. With the projection of meaning to the "exterior world" effectively withdrawn, a "phenomenal residue" (Lauer, 1965, p. 27) of immanent givenness will remain. Following the phase of *epochē*, one is to undertake "the more positive technique of 'reduction,' whereby the phenomenal residue is rendered more and more emphatically present to consciousness" (p. 27). Overall, the intended effect of the phenomenological method is to make consciousness more self-reflective; increasingly, it is to refer to itself alone, to its own innate activity. Under naturalism, reference to the "external world" narrows the truth into an object-in-space and denies the subjective source of this truth; the truth is thereby limited to mere existential particulars. In contrast, phenomenological *self*-reference permits a direct insight into the subject's very act of intentional object-making. Known in the immediacy of intuitive self-reflection, this activity is adequate unto itself and requires no "empirical validation."

For Husserl, the apodictic action of self-consciousness entails pure rationality. The knowledge it affords, being unadulterated by the contingencies of empirical fact, is essential; it is both necessary and universal. As Lauer puts it,

> [A]bsolute givenness and hence absolute being are available to a careful reflection on the acts of consciousness. In the act itself is revealed the manner in which the intentional object of the act is given, which is to say that the act contains its own evidence, its own guarantee of givenness. (1965, p. 60)

It is clear from Husserl's *Logical Investigations*, and from his later work, *Formal and Transcendental Logic*, that the "guarantee of givenness" that he sought must come from a "necessity that is logical rather than psychological" (Lauer, 1965, p. 61). "Essential intuition" then, as an internal reflection on operations that are necessary and universal, must be an intuition of that which is logical through and through. Yet Lauer notes a change in Husserl's approach from the publication of *Logical Investigations* in 1900 to the appearance of *Formal and Transcendental Logic* in 1929. Logical truth becomes "progressively 'transcendentalized'" so that now,

> ideal truth [is] more consciously rooted in pure consciousness, or to use the later terminology, in the universal a priori of transcendental subjec-

> tivity. With *Formale und transzendentale Logik* phenomenology has become more radically subjective, but in so doing, thinks Husserl, it has become more radically objective. Being, truth, and evidence are now thoroughly rooted in phenomenological consciousness.... (Lauer, 1965, p. 58)

In other words, with Husserl's move toward subjectivity *as such*, he finds an *ontological* basis for grounding logical truth. Said truth, in its absolute being, is none other than pure subjectivity. This equation of being with knowing is obviously reminiscent of Descartes's *cogito ergo sum*. Lauer says accordingly that, "though one can discern in transcendental phenomenology a theory of being and a theory of truth, these are in reality indistinguishable from its theory of the validity of cognition, which in its turn is ultimately a theory of evidence. Absolute being is the object of a cognition that is absolutely true" (1965, p. 47). Therefore, when Lauer comments that, for Husserl, "God is replaced by 'absolute being'" (p. 47), in fact he means that God is replaced by the *cogito*—something that Descartes, for his part, was as yet unable to do. What this amounts to is a Nietzschean enthronement of the *cogito* or "transcendental ego" (Husserl, 1936/1970) in the wake of "God's death." Perfected through the achievement of a pure reflection upon its own inherent activity, the ego becomes totally adequate unto itself. It has now overcome all dependence on the "outer world" of empirical facts. In surmounting all the uncertainties of existential life, all the ambiguities of *apeiron*, the transcendental ego reaches its "mountaintop" and dons the mantle of the overman. This is Husserl's modernist Faustian vision, at any rate.

Although Husserl's thinking did evolve during the first three decades of the twentieth century, in a more basic way perhaps, it remained the same. Lauer, Husserl's translator and editor, makes it easy for us to see this by juxtaposing Husserl's 1911 essay, "Philosophy as Rigorous Science," with a much later work, his 1935 lecture on the "Crisis of European Man." (This lecture was an introduction to Husserl's *Crisis of the European Sciences and Transcendental Phenomenology* [1936/1970].) According to Lauer, these two essays, taken together, "constitute a striking testimony to the continuity of Husserl's 'scientific' ideal in philosophy" (1965, p. 2).

Examining the later presentation, we find a restatement of Husserl's by-then familiar critique of naturalistic objectivism, this time seen in historical perspective. Husserl thus speaks of the "naïveté...characteristic of philosophy in the whole modern period since the Renaissance....[T]he most general title for this naïveté is objectivism, which is given a structure in the various types of naturalism, wherein the spirit [that is, the psyche or consciousness] is naturalized. Old and new philosophies were and remain naïvely objectivistic" (Husserl, 1935/1965, p. 181). Husserl goes on to denounce once again the doctrine of external causality, found in both physics and psychology: "In the light of psychophysical causality there is no purely spiritual continuity in spiritual grouping; physical nature envelops everything" (p. 182). As a consequence,

> There can be no pure, self-contained search for an explanation of the spiritual, no purely inner-oriented psychology or theory of spirit beginning with the ego in psychical self-experience....The way that must be traveled is the external one, the path of physics and chemistry....[T]his objectivism or this psychophysical interpretation of the world, despite its seeming self-evidence, is a naïve one-sidedness that never was understood to be such. (p. 184)

Therefore, since "no objective science can do justice to the subjectivity that achieves science" (p. 185), what is required is a science of consciousness that can liberate itself from naturalistic objectivism and function strictly within "the closed circle of its own knowing activities" (p. 171), a science involving "the highest level of reflection upon itself" (p. 181). In short, a *phenomenological* science is required, for "only if the spirit returns to itself from its naïve exteriorization, clinging to itself and purely to itself, can it be adequate to itself" (p. 189). Husserl concludes in the following way:

> I do hope...to have shown that we are not renewing here the old rationalism, which was an absurd naturalism, utterly incapable of grasping the problems of spirit that concern us most. The *ratio* now in question is none other than spirit understanding itself in a really universal, really radical manner, in the form of a science whose scope is universal, wherein an entirely new scientific thinking is established in which every conceivable question, whether of being, of norm, or of so-called "existence," finds its place. It is my conviction that intentional phenomenology has for the first time made spirit as spirit the field of systematic, scientific experience, thus effecting a total transformation of the task of knowledge. (pp. 190–191)

Now, I am contending that Husserlian philosophy is a prime example of *modernism*—the cultural movement implicitly dedicated to the attempt to deny *apeiron* in the face of her mid-nineteenth century resurgence. At all costs, the modernist seeks to maintain that old formula, *object-in-space-before-subject*. But was Husserl not *attacking* objectivism? In actuality, what he was denouncing was only the classical form of objectivism, that in which objects were projected into an "exterior world of nature." Husserl recognized the bankruptcy of this mechanistic approach more clearly than did many of his contemporaries. For, while the empirical science of the day did seem to be heeding the message that it was time to start "looking within"—as witnessed by the advent of scientific psychology—it could not break the habit of turning whatever it looked at into an object taken to exist in the world "out there." One result of this was the prevalence of the reductionistic view that mental states are merely epiphenomena of the nervous system. But even when no specific assumption was made that the mind "resides" in the natural object called the "brain," it was still treated as if it were somehow "out there," detached from the consciousness of the investigator, so that introspectionist psychology could convince itself that it was studying the mind in an "empirical" fashion. What Husserl did was insist that consciousness be treated as strictly *internal*, as an object completely immanent to the investigator's own consciousness. Yet it was an object nonetheless.

As Lauer puts it, scientific phenomenology entails an "immanent objectivity....A philosophy so conceived gradually reveals the essence of its own object, and in so doing, it progressively modifies itself in accord with this object, since it is the object, not the subject, that determines what the science must be" (1965, pp. 22–23). The statement is further clarified in a footnote: "It is the character of what is to be known, not the character of the knower, that makes phenomenology the only viable philosophical method" (p. 23, n. 28). Having rejected empirical objectivism, why must Husserl commit himself to an *immanent* objectivism? It is simply because Husserl's rationalism compels him to seek the fixity of "essential truth," of that which is both necessary and universal. This in turn requires a truth that is completely objective.

In Part One, we found that the establishment of scientific objectivity depends on the object being secured within a framework that permits complete access to it, thereby allowing it to be known without ambiguity. The framework necessary for the scientific observation and analysis of the object is the *spatial continuum*. To be sure, in rejecting *empirical* objectivism, Husserl sets aside its exterior space. In its place is a wholly interior dimension, one that is to be known through intuition rather than sense perception; one that is neither physical nor empirically psychological but purely logical —a space without loopholes (the logical equivalent of "black holes"!) wherein the perfectly self-consistent operations of consciousness are enacted. Of course, it is consciousness itself that is to observe these operations. Therefore, the "observation" in question is a "*self*-observation," an intuitive self-reflection. And yet, the old categorial division of subject and object can still be found in this immanent domain. For, to know anything objectively, the knower or subject must detach himself from what he knows. What we saw in Part One is that, in scientific objectivism, the containment of an object in space and the detachment of the subject from that object-in-space are integral aspects of the same process: by the same act in which the subject seals its object into space, it seals itself out, separates itself from that object. No less does this apply to Husserl's phenomenological version of scientific objectivism. If the "transcendental subject" is to gain purely apodictic knowledge of its object—which, in this case, is itself—it must become detached from that object. The internal division of consciousness into subject-as-object and subject-as-subject is implied in Lauer's comment that, for Husserlian phenomenology, "it is the object, not the subject, that determines what science must be," and that it "is the character of what is to be known, not the character of the knower, that makes phenomenology the only viable philosophical method" (1965, p. 23).

Note the limitation of Husserlian intuition that results from its objectivism. It cannot take in the act by which it splits itself into subject-as-object and subject-as-subject. Its grasp is restricted to the object alone, which appears only after the division has occurred. The division itself is thus not open to intuitive inquiry but is presupposed unquestioningly. Husserlian reflection confines itself to consciousness as a *content*, as the "objective action" that is contained in the interior logical space that has been established. It is in this way that the phenomenological *cogito* is divided, with one part of it—the part to be investigated—constituting an "interior logical object," and the other part of it—that doing the investigating—entailing a new, higher-order subjectivity that goes uninvestigated. The end result of this is the establishment of a regress that guarantees the failure of Husserl's "phenomenological reduction." For, in the attempt to reach "objective closure" on the "transcendental subject" via the division of consciousness, subjectivity forever evades the grasp of intuition.

Now, it is *logical* closure that Husserl sought, and the ultimate defeat of his project has an interesting parallel. In the last decade of Husserl's career, the logician Kurt Gödel dealt a devastating blow to mathematical logic. Included in his momentous 1931 findings is the theorem that states that any mathematical system extensive enough to embrace the natural numbers is incapable of proving its own logical consistency. The consistency of the system *could* be established, but this would require appealing to a *more comprehensive* mathematical system, which, in turn, would be faced with the problem of proving *its* own consistency, and so on, ad infinitum. In his attempt to surpass naturalistic rationalism, Husserl, in effect, was "appealing to a more comprehensive logical system." When he asserted that, "no objective science can do justice to the subjectivity that achieves science" (p. 185), he was referring to the classical science that excludes subjectivity from its account, either by "naturalizing" or "empiricizing" it, or simply by ignoring it. Husserl hoped to remedy this by expanding science so as to include what had erroneously been left out. But science was still to be strictly objective, which meant that Husserl had to perform an exclusion of subjectivity in his own right. Therefore, the statement that he directed against the naturalistic brand of objective science is equally applicable to his phenomenological attempt at scientific objectivity. Because Husserl also had to drop subjectivity from consideration (albeit at a higher level of abstraction), he was unable "to do justice to the subjectivity that achieves science." It was this that prevented objective closure on logical subjectivity, and this that pointed to the need for appealing to a still more comprehensive, more abstractly inclusive system of self-objectification, which alas would require a further subdivision of consciousness. As long as the aim was logical closure—the perfected self-consistency of apodictic reason, the process could never end. Abstraction would be heaped upon abstraction with no true closure ever really being gained. The phenomenological division of the classical subject into a "subject-as-object" and a "subject-as-subject" thus would be reapplied to the latter to produce a "(subject-as-subject)-as-object" and a "(subject-as-subject)-as-subject." The reapplications would continue in this way with no final resolution. At bottom, the motive for embarking on such a course is the desire to banish *apeiron* once and for all and arrive at the pure rationality of the transcendental ego. But this quest for egoic unity is a chimera and the ego only winds up dividing itself again and again, fragmenting itself into ever more vaporous self-reproductions. So the individual—instead of achieving individuation, indivisible wholeness—is ceaselessly divided. The "atom" is split. ("Atom" is from the Greek, *a-tom*, not divisible.)

The failure of Husserl's project mirrors the failure of modernism in general, as described in Part One. Without expressly stating it, Einstein, Picasso, Freud, and every other modernist had been essentially concerned with accommodating resurgent *apeiron* in such a way as to deny it. This was to be accomplished by objectifying the subject—something Nietzsche's early modernism seemed less able to achieve than the forms of modernism that came later. But what we have now seen in examining Husserlian philosophy—and what we found in Part One for the fields of physics, art, and the media—is that the success of modernism was actually never as real as it was apparent. Once again, the crux of the problem is that modernism's hope of gaining closure by objectifying the subject is implicitly frustrated by the opening up of a new, more abstract order of subjectivity that itself is *not* objectified. Because modernist objectification necessarily prefigures a "re-subjectification," it defeats its own purpose and there is a collapse of meaning. It was modernism that was to compensate us for the "death of God"—for the demise of the old system of external meanings and values—by relocating meaning within, centering it within the *cogito*. But, by the third and fourth decades of the twentieth century, there was a nagging suspicion that the *cogito* itself was on its deathbed.

The hardest thing to accept is one's own death. Therefore, even when death is staring one in the face, one is likely to go on denying it. By and large, this was the initial reaction to the impending demise of the modernist *cogito*. The "re-subjectification" of the *cogito* was ignored and a brave face put on. "All will prove objective in the end," the modernist groundlessly assured himself. In fact, this "up-beat" attitude is still maintained by much of modernist science. Yet, as the twentieth century progressed, the bankruptcy of modernism came closer and closer to the threshold of recognition in other spheres of contemporary culture. A *post*modern mood began to take hold. The distinguishing features of postmodernism were recounted in Part One. One can no longer effectively deny the loss of objective meaning, yet neither can one really accept it, since one's very life—the life of the *cogito*, that is—seems to depend on maintaining objectivity. The reaction to this dilemma is *regression*: One enters an endless regress in which each frustrated attempt to achieve objective meaning is followed by another such attempt, and another. Of course, logically, the "meaning addict" knows that the mere repetition of an ineffective action does not make it effective. But logic of the cerebral kind no longer applies here, given that the addict's very survival appears to be on the line. In the addict's more primal "logic," it really doesn't matter that every effort at affirmation ends in failure. Because no other alternative appears even thinkable, all that matters is the momentary ability to keep on making the effort. In fact, this pattern of repetition can become a kind of game on which the addict may get "high," as we shall soon see.

I want to emphasize that modernism and postmodernism are not opposing cultural phenomena, nor are they even separate. In that modernism's defining act of self-objectification at once implies its "re-subjectification," the seeds of postmodernism are planted in it from the very beginning. Therefore, it is not surprising that modernist philosophy should eventuate in a postmodern alter-philosophy, much in the same way that art and media have transmuted themselves (and science as well, though perhaps more reluctantly and in a less obvious manner). It was in the post-World War Two milieu of the 1950s and 1960s—when television was coming into its own as the premier postmodern medium, and when the artworld was undergoing its own transformation through the works of Jasper Johns and Andy Warhol—that postmodern philosophy was emerging. The prime exemplar of this movement is the French deconstructionist Jacques Derrida (b. 1930).

4. *Différance*

By "deconstruction," Derrida does not mean "demolition," but "de-sedimentation" (1976, p. 10). In the fashion of an archeologist, one is to drill down into and through the accumulated layers of Western metaphysics until one reaches its origin (*archē*). What foundation stone of philosophy do we uncover through such a deconstruction? It is the Greek *phōnē*, the voice or sound. To hear "oneself" speak is to experience the "one in here," the self-identical speaker before whom is projected a "world out there." Thus Derrida says:

> The system of "hearing (understanding)-oneself-speak" through the phonic substance—which *presents itself* as the nonexterior, nonmundane, therefore nonempirical or noncontingent signifier—has necessarily dominated the history of the world during an entire epoch, and has even produced the idea of the world, the idea of world-origin, that arises from the difference between the worldly and the non-worldly, the outside and the inside, ideality and nonideality, universal and nonuniversal, transcendental and empirical, etc. (1976, pp. 7–8)

Derrida stresses "the original and essential link to the *phonè*" of the *logos*: "the essence of the *phonè* would be immediately proximate to that which within 'thought' as logos relates to 'meaning,' produces it, receives it, speaks it, 'composes' it" (1976, p. 11). Thus, "for Aristotle…'spoken words…are the symbols of mental experience'" (Derrida, 1976, p. 11). Writing in the same vein, Derrida alludes to the "heritage of that logocentrism which is also a phonocentrism: absolute proximity of voice and being, of voice and the meaning of being, of voice and the ideality of meaning" (pp. 11–12). Next—after citing G. W. F. Hegel's demonstration of "the strange privilege of sound" in producing "the self-presence of the subject"—Derrida makes phono/logocentrism responsible for "the historical determination of the meaning of being in general as *presence*" (p. 12). This includes: "presence of the thing to the sight as *eidos*, presence as substance/essence/existence, temporal presence as point of the now or of the moment, the self-presence of the cogito, consciousness, subjectivity, the co-presence of the other and of the self, intersubjectivity as the intentional phenomenon of the ego, and so forth" (p. 12). In short, the primacy of the spoken word is intimately bound to the primacy of an "interior" *cogito*, and these give rise to a "philosophy of presence" so pervasive in Western thought that Derrida feels able to equate it with "philosophy *itself*" (p. 12). It is this that he wishes to deconstruct.

Having stripped away the sediments and arrived at the core of Western language and thought—namely, phonologocentric self-presence, does Derrida simply confirm this as the bedrock of being, its absolute foundation and irreproachable origin? On the contrary, he announces that—notwithstanding the "intuitive self-evidence" of an entire epoch of human history, the world in fact does *not* begin with the self-identity of the spoken word. Does it originate then from a purely *non*-linguistic source? Does it stem from some "instinctual essence" in which there is no symbolic or representational expression of any kind? Not at all. Instead of the world arising from a primordial substratum characterized by the absence of language, it originates from a *language of absence*. Derrida calls this language "primary writing" (1976, p. 7) or "arche-writing" (pp. 56–57, 60, 68).

Writing? Does the written word not derive from the spoken word? Is writing not merely an adjunct to speech, a practical technique for preserving more permanently than can memory the "interior presence" of the "true" word by inscribing it upon an exterior surface (the writing tablet)? So it would seem if one holds with the tradition of phonocentrism. Thus, Aristotle, after asserting that "'spoken words are the symbols of mental experience,'" added that "'written words are the symbols of spoken words'" (Derrida, 1976, p. 11). And two thousand years later, Ferdinand de Saussure, a founder of modern(ist) linguistics, could still say: "'Language and writing are two distinct systems of signs; the second *exists for the sole purpose of representing* the first'" (p. 30; Derrida's italics). In this way, "the epoch of the logos...debases writing considered as mediation of mediation and as a fall into the exteriority of meaning" (pp. 12–13). But, according to Derrida, the rule of *phōnē* and *logos* is now being brought into question, and, in its deconstruction, the primary role of writing is beginning to be disclosed.

We should not expect that "primary writing" should look familiar to us, given the fact that writing as we have known it has been constrained so tightly, for so many centuries, by logocentric speech. Indeed, in paying heed to the movement of arche-writing, we destabilize and de-realize all the parameters of writing that have previously operated. Whereas phonocentric writing is but the outward expression of the presence of the "inner voice," arche-writing is the *absence* from which this presence itself arises. No longer are we to take as simply self-originating the stability of the self-identical that is anchored in "hearing-oneself-speak." Instead, we must situate this identity in a field of *difference*, "a structure of reference where difference appears *as such* and thus permits a certain liberty of variations among the full terms" (Derrida, 1976, pp. 46–47). In the hyper-animated, forever self-effacing milieu of original writing, "the completely other is announced as such—without any simplicity, any identity, any resemblance or continuity" (p. 47). Here, instead of omnipresence—an "intuition of a present truth [that] stretches out to give [writing a] foundation" (p. 48)—there is the "nonintuition" (p. 40) of an *omni-absence*, a restless interchange of signs preemptively engaged in the act of displacing each other. "From the moment that there is meaning there are nothing but signs. We *think only in signs*. Which amounts to ruining the notion of the sign" (p. 50). For the process is never brought to order by securing a "transcendental signified" once and for all:

> There is...no phenomenality reducing the sign or the representer so that the thing signified may be allowed to glow finally in the luminosity of its presence. The so-called "thing itself" is always already a *representamen* shielded from the simplicity of intuitive evidence. The *representamen* functions only by giving rise to an *interpretant* that itself becomes a sign and so on to infinity. The self-identity of the signified conceals itself unceasingly and is always on the move. The property of the *representamen* is to be itself and another, to be produced as a structure of reference, to be separated from itself. (Derrida, 1976, pp. 49–50)

Derrida characterizes the incessant displacements, deferrals, decenterings and discontinuities of original writing (*—les brisures* [p. 65]: hinges, cracks, crevices, *holes—*) as *play*: "This *play*, thought as absence of the transcendental signified, is not a play *in the world*, as it has always been defined, for the purposes of *containing* it." Instead of being a game *in* the world, it is "the game *of* the world" (my italics). It is this "that must first be thought, before attempting to understand all the forms of play in the world" (p. 50).

Crucial to Derrida's notion of the archegraphic play of differences ("*différance*") is the idea of the "trace." His translator explains that this "French word carries strong implications of track, footprint, imprint...[and therefore] presents itself as the mark of an anterior presence, origin, master" (Spivak, 1976, p. xv). It is clear that the trace itself "cannot be a master-word" (p. xv), for it is only a surface effect. Translator Gayatri Spivak observes that, "For 'trace' one can substitute 'arche-writing'...or 'différance'" (p. xv). Original writing then, entails the interplay of epiphenomenal traces. The logocentric expectation is that primary causes underlie such secondary effects. For is it not *intuitively obvious* that, if a footprint is to be left, the anterior presence of a foot is required? Yes, this is surely a "self-evident truth" within the old metaphysics of presence, but it is precisely such an intuition that Derrida wishes to disrupt. Accordingly, we are to grasp the play of traces in arche-writing as entailing aftereffects without antecedents. No "feet" precede these strange "footprints." Like the ghost of a being that was never alive, the trace is wholly ephemeral. As pure play, the game of the written trace is weightless, devoid of the "gravity of speech" (p. 50). A principle of *levity* prevails here that brings to mind the smile of the Cheshire cat, or better perhaps, the mad laughter of the shepherd in Nietzsche's *Zarathustra*. (Indeed, Derrida acknowledges that "Nietzsche...contributed a great deal to the liberation of the signifier from its dependence or derivation with respect to the logos"; 1976, p. 19.) Traces are bubbly, effervescent; they ceaselessly burst and reform in "mid-air." Nothing more sober lends ballast to this levity; there is nothing below the trace to provide it with a solid foundation; no sub-stantial presence to serve as its ground; no substratum to actualize its virtuality.

Arche-writing occurs upon a surface without depth. Or, to be more faithful to Derrida, the dimension of the trace is a "horizontality" (1976, p. 69) in which there is not even any solid distinction between surface and depth, exteriority and interiority. It is within the phonocentric tradition that "it must be believed that there is an *inside* of…language" (p. 33) set in opposition to what is outside. In fact, phonocentrism, logocentrism, or any other centrism *must* posit this kind of hard distinction, since a center is what it is by categorically differentiating itself from a periphery, from that which lies outside of it (Derrida thus opens his book by characterizing logocentrism as "nothing but the most original and powerful ethnocentrism"; p. 3). But arche-writing entails a "relationship that can no longer be thought within the simple difference…of 'image' and 'reality,' of 'outside' and 'inside,' of 'appearance' and 'essence,' with the entire system of oppositions which necessarily follows from it" (p. 33). The notion of the trace—*as* an "*arche-*writing"—points to something more "fundamental" than this; something "anterior" to the concept of inside and outside, and to all the metaphysical dualisms associated with it. Preceding the depth of phonologocentrism is the "horizontal" movement of *différance*, the field of pure activity and play that indeed first gives rise to inside and outside, but only with a smile or a laugh. "Constituting and dislocating [this difference] at the same time" (p. 68), there is the implication that it is just "a simple *moment of the discourse*" (p. 61), as are all differences. In the "originary" field of pure difference, of difference *as such* (*différance*), particular differences themselves are rendered "different"; that is, they are thoroughly relativized.

Perhaps it is obvious why I have bracketed the words "originary," "fundamental," and "anterior." The notion of priority, of "coming first," is the very thing that Derrida wishes to deconstruct with his concept of *différance.* Then, in speaking of "*arche-*" or "*original* writing," does he not undermine his own anti-foundational program by implying a new foundation in place of the old? Derrida is well aware of the problem:

> The trace is not only the disappearance of origin…it means that the origin did not even disappear, that it was never constituted except reciprocally by a nonorigin, the trace, which thus becomes the origin of the origin….[T]o wrench the concept of the trace from the classical scheme, which would derive it from a presence or from an originary nontrace and which would make of it an empirical mark, one must indeed speak of an originary trace or arche-trace. Yet we know that that

> concept destroys its name and that, if all begins with the trace, there is above all no originary trace. (1976, p. 61)

Derrida is giving us an example here of what he calls "the erasure of concepts" (p. 61). The concept of the "arche-trace" is first posited within the classical context as if it actually could *be* a simple origin, whereupon we notice the contradiction of such an "originary nonorigin," and this leads us quickly to negate what we have posited: if the trace is the origin, there *is no* origin. Thus the notion of the arche-trace is put under erasure (*sous rature*): it is affirmed, only to be negated. Derrida is not suggesting here that the negation of presence achieves any permanence. This is reflected in the fact that the technique of Derridean erasure does not simply eradicate the term of presence ("origin," for example) so that it is no longer visible. As Derrida's translator explains, we "write a word, cross it out, and then print both word and deletion" (Spivak, 1976, p. xiv). We do this because, though the word is inaccurate, it also remains *necessary*: "Since the word is inaccurate, it is crossed out. Since it is necessary, it remains legible" (p. xiv). Herein lies the basic movement of Derridean thought. We go from the word to its erasure, then slip back into the word (since it continues to be "necessary"), erase that word once more (since it is "inaccurate"), and so on, indefinitely. A vicious circle is thus created in which each denial is itself denied, only to bring on the next: presence→absence→presence→absence→presence... While the action never comes to rest on either term, terms are neither neutralized nor integrated. Instead they remain as they always have, though now in a relation of ambiguity, since the term of absence no longer stays merely implicit.

It is clear that, with the work of Derrida, we enter the infinite regress of postmodernism prefigured by modernism itself:

> In this play of representation, the point of origin becomes ungraspable. There are things like reflecting pools, and images, an infinite reference from one to the other, but no longer a source, a spring. There is no longer a simple origin. For what is reflected is split *in itself* and not only as an addition to itself of its image. The reflection, the image, the double, splits what it doubles. The origin of the speculation becomes a difference. What can look at itself is not one; and the law of the addition of the origin to its representation, of the thing to its image, is that one plus one makes at least three. (1976, p. 36)

What Derrida describes is de-individuation: an unconstrained "splitting of the atom" that disrupts every center, a "nuclear meltdown" through the "containment vessel" of classical metaphysics. This schismatic dissociation, this world-schizophrenia that Derrida calls the "*game* of the world," is hardly to be read as *transcendence*. On the contrary, it is to be seen as the recognition of the necessary failure of all such attempts at transcending metaphysics, since transcendence is itself the chief project of metaphysics. So "difference *as such*," in undermining all merely relative differences (presence vs. absence, positive vs. negative, inside vs. outside, etc.), undermines *itself* as something different from what is relative. Relativity alone prevails.

In his critique of Husserl, Derrida retraces the pathway that leads inescapably from modernism to postmodernity. He does acknowledge as "indispensable" the "Husserlian correction" of the "prephenomenological psychology of the imagination" in which the psyche is designated as "*another natural reality, internal and not external*" (1976, p. 64): "The psychic image…must not be an internal reality copying an external one. Husserl, who criticizes this concept of 'portrait' in *Ideen I* shows also in the *Krisis*… how phenomenology should overcome the naturalistic opposition—whereby psychology and the other sciences of man survive—between 'internal' and 'external' experience" (Derrida, 1976, p. 64). We have seen that Husserl attempted to supersede naturalism by the quintessentially modernist move of bringing consciousness to closure as the intuitive transparency of *self*-consciousness. And we also have witnessed the *failure* of this self-objectification brought on by the concomitant *re-subjectification* that opens the doorway to postmodernism's "hall of mirrors." Derrida points out that Husserl, in clinging to the metaphysics of presence, interprets "this [post-modern] nonintuition teleologically as *crisis*," since it is "for Husserlian intuitionism that which exiles us far from the *clear* evidence of the sense, that is to say from the full presence of the signified in its truth" (p. 40). But this "crisis" is evidently of Husserl's own making, for it is "linked with the very moment of truth and the production of ideal objectivity" (p. 40). Therefore, rather than taking the loss of objectivity as if it were a "mere accident" (p. 40) that could be overcome by properly addressing the "crisis" it has brought, Derrida suggests that "it is…the concept of crisis that should be suspect, by virtue of what ties it to a dialectical and teleological determination of negativity" (p. 40). In other words, if the counter-intuitive mirror-play of difference is not merely an adventitious development but implicit in Husserl from the outset, then it surely does not constitute a crisis in the sense of a negativity that has intruded itself from an outside source and that is thus susceptible to being expelled. Instead, because the "crisis" of the ever-vanishing trace is actually *integral* to Husserl's formulation, we must conclude that it has always been there and always will be, that it is *originary*. Accordingly, right after commending Husserl for his helpful critique of naturalism, Derrida steps beyond Husserl's "crisis mentality" with respect to the trace by reminding us once again of its originary status—not only in Husserlian metaphysics, but also in *all* metaphysics. The arche-trace is:

> absolutely and by rights "anterior" to all...metaphysical problematics concerning the meaning of absolute presence whose trace is thus opened to deciphering. *The trace is in fact the absolute origin of sense in general. Which amounts to saying once again that there is no absolute origin of sense in general. The trace is the différance* which opens appearance and signification....[The] origin of ideality, the trace is not more ideal than real, not more intelligible than sensible, not more a transparent signification than an opaque energy and *no concept of metaphysics can describe it*. (1976, p. 65)

So much, then, for Husserl's notion of the intuitive transparency of consciousness. Derrida's very manner of questioning Husserl is in fact a testimony to the play of *un*conscious forces, for in it we see how words forever slip away from themselves into the open-ended ambiguity of abstract self-mirroring. Indeed, Derrida explicitly states that the structure of the signifier operates "according to the causality of the unconscious" (p. 89), and he pays tribute to psychoanalysis (p. 88), and, in particular, to Freud (p. 67).

Now, Derrida registers a certain "optimism" about the liberation of writing from the oppressive rule of logocentric speech. His attitude toward the prospect of deconstructing metaphysical positivism is curiously "positive":

> By a slow movement whose necessity is hardly perceptible, everything that for at least some twenty centuries tended toward and finally succeeded in being gathered under the name of language is beginning to let itself be transferred to...the name of writing....There is not a single signified that escapes...the play of signifying references that constitute language. The advent of writing is the advent of this play; today such a play is coming into its own....It is as if the Western concept of language...were revealed today as the guise or disguise of a primary writing. (1976, pp. 6–7)

Among Derrida's prime examples of the contemporary emergence of "primary writing" is *cybernetics*, with its *pro-gram*: "the entire field covered by the cybernetic *program* will be the field of writing. If the theory of cybernetics is by itself to oust all metaphysical concepts...it must conserve the notion of writing, trace, grammè [written mark]" (p. 9). Without yet spelling out the unique nature of cybernetic writing, Derrida next proceeds to reinforce his claim that "a writing thus enlarged and radicalized...inaugurates the...deconstruction" (p. 10) of Western philosophy. Later on in his text, Derrida does attempt to provide more detail about cybernetics. To prepare for his rendition of cyber-writing, he explicitly associates it with his notion of *différance*, with the newly emergent idea of the ceaseless movement of the trace: "This movement goes far beyond the possibilities of the 'intentional consciousness' [thus beyond Husserlian phenomenology]. It is an emergence that makes the *grammè* appear *as such* (that is to say according to a new structure of nonpresence)" (p. 84). We then learn that computer innovations such as "electronic card-indexes and reading machines" (p. 84) are currently breaking the hold of the old linear writing—writing that has been subordinated to the phonocentric "ideal of continuous movement, straight or circular" (p. 86). The novel cybernetic writing is decidedly *non*linear; it is discontinuous or discrete (today we say *digital*); it is "pluri-dimensional" and thus is "not subjected to successivity, to the order of a logical time, or to the irreversible temporality of sound" (p. 85). Moreover, the "pluri-dimensional symbolic structure is not given within the category of the simultaneous. Simultaneity coordinates two absolute presents, two points or instants of presence, and it remains a linearist concept" (p. 85). Nevertheless, from Derrida's citation of André Leroi-Gourhan's prophetic 1965 anticipation of things to come, we can see that cyber-writing has its own kind of "simultaneity":

> "Scientific thought is...hampered by the necessity of drawing itself out in typographic channels and it is certain that if some procedure would permit the presentation of books in such a way that the materials of the different chapters are presented simultaneously in all their aspects, authors and their users would find a considerable advantage....A vast 'tape-library' with an electronic selection system will in the near future show pre-selected and instantaneously retrieved information." (1976, pp. 332–333)

The cyberspace "simultaneity" apprehended by Leroi-Gourhan must not be confused with that of *classical* space. Instead of constituting the actuality of what simply is co-present, it is a *potential* simultaneity, one in which information is not really "presented" (Leroi-Gourhan's questionable terminology notwithstanding), but is held in reserve for the *possibility* of presentation and use. This does not mean that the information reserve, while not being immediately present *to view*, is palpably present *elsewhere* in the computer; that, in this tangible form, it lies "somewhere inside the machine" just waiting to be drawn upon. Let us try to understand better exactly what it does mean.

Part One of this book is largely concerned with the historical transition from the mechanical age to the electronic age or Age of Light. Today, we are more likely to describe our electronic epoch as the Age of *Information*. Whereas classical machines entail palpable operations upon matter and energy, computers involve the subtler activity of information. To illustrate the essential difference, let me compare the word processing that I now am doing with the production of words on a mechanical typewriter.

The old typewriter is fitted with a fixed set of type bars, each with its own alphabetic or numeric character. The typefaces are actuated by stroking keys on a keyboard composed of corresponding characters. If, for example, I press the key marked "n," mechanical energy is imparted to the associated type bar causing it to rise out of its housing and strike an inked ribbon secured over a blank sheet of paper that is held in place on a cylindrical platen. In this way, the metallic typeface, "n," is made to leave its trace on the page. What process produces the letters now appearing before me on this computer display?

The close resemblance of my computer keyboard to that of the mechanical typewriter is deceptive. When I tap the "n" key on my word processor, I do not initiate a mechanical operation wherein an already present typeface is *re*-presented, transferred to a blank surface. Instead, I send an electronic impulse to a microprocessor that, in itself, possesses no such preexisting actualization of the letter "n." What we find in the hardware of the computer, in the silicon chip in which the computer's operations are carried out, are not so much material actualities as *informational potentials*. Physicist David Bohm explains: "in a computer, the information in a particular chip has a wide range of virtual or potential activities to which it may give rise. Only some of these are actualized in the activity of the computer as a whole, in a way determined by the overall context of the entire structure of the computer and by all the information that has been put into it" (1986a, p. 125). According to Bohm, information is defined as "a *form* that literally 'informs' (i.e., forms from within) an 'unformed' energy to give rise to a corresponding determinate activity" (p. 126). And with respect to *computer* information, "the form in the state of the silicon chips enters into the energy in the computer to 'give shape' to a corresponding activity" (p. 126). The chip is composed of a large array of tiny transistor switches each of which constitutes a basic unit (or "bit") of information in that it can answer the simple question of whether the switch is on or off, one or zero (more exactly, whether or not voltage is applied to the silicon diode in such a way that the flow of electrons is permitted). When I give the chip input by pressing the buttons on my keyboard, it responds to me by "flipping switches," choosing zeroes and ones, in accordance with the way it has been programmed. In the specific example, when I type the letter "n," the microchip creates a particular binary pattern of zeroes and ones as per the instructions it receives from the word processing program that I am using. It then transforms this pattern into a recognizable output that appears on the monitor as a visual display: "n." In this way, informational potential is actualized, activity is shaped that leads to the creation of meaning. Thus, whereas mechanical typewriting entails the transfer of an *already* actualized pattern of matter-energy (the typefaces) from one location in space to another (from the bank of type bars to the sheet of paper), electronic word processing involves an actualization *in-the-making* whereby a pattern of matter-energy is produced from an initially potential state, one in which the pattern does not yet exist. Since bits of information are not committed in advance to any particular combinatorial pattern, and since zeroes and ones can be combined in an indefinite variety of ways, meaning can be shaped in a highly flexible manner. For example, we are not limited to a fixed set of characters in word processing but can create any number of character sets and switch from one to another at the press of a button. The flexibility of word processing is greatly enhanced by the fact that the electronic meaning patterns appearing on the computer display are *virtual*. The patterns are not *fully* actualized until a final printout of hard copy is staged, and this permits the text to be sculpted in a far more fluid manner than is possible in typewriting or handwriting.

What it comes down to in Derrida's terms is that mechanical writing is governed by the metaphysics of presence whereas cyber-writing is not. The typed or handwritten character is a squiggle of ink or graphite that functions as a substantive reality. Matter concretized in this fashion is certainly not immutable; it can be altered in a variety of ways via exchanges of energy. Nevertheless, the metaphysics of presence is enforced here by the first law of thermodynamics, which tells us that—however matter-energy might be continuously transformed—it cannot simply *vanish* from the space-time continuum; it can neither be destroyed nor created, we are told. Accordingly, the printed mark—as a material presence fixed upon the writing surface by the chemicals in its ink or graphite—cannot be eradicated without at least leaving some *trace*. This trace—say, in the form of a faint darkening of the page where a letter was erased, or a slight unevenness of the surface where correction fluid was applied, or perhaps in a form so subtle that a microscope would be required to detect it—attests to the fact that, although the sign has indeed been transformed in the attempt to erase it, it never completely disappears from the page.

But would the printed mark really need to be annihilated from space-time in order for it to be removed from the writing surface without a trace? In mechanical writing, the initial inscription of the sign is obviously no creation *ex nihilo*; it is a transfer of energy from a preexistent source (typeface, pen, pencil, etc.). Therefore, even if the first law of thermodynamics cannot be violated by destroying the printed mark altogether, could the mark not at least be transformed in such a way that it would revert to the form of energy in which it existed prior to the act of inscription? Would not such a reversal effectively remove all traces of the inscription that had appeared on the writing surface? This is where the *second* law of thermodynamics enters the picture. According to the principle of entropy, the transformation of matter-energy from one state to another in fact *cannot* entirely be reversed. For instance, in burning a lump of coal, potential energy is converted into kinetic energy in such a way that an irreducible residue is left; the ashen byproduct of the transaction bears witness to the fact that the potential originally contained in the coal cannot be restored. In a similar way, once a sign is hard-printed onto a page, there is no going back to the state of affairs that existed prior to the inscription. This inability to totally reverse the engravement process affirms the "gravity" of such writing. Because mechanical writing bears the weight of substantival presence, the traces it leaves cannot fully be effaced.

It is evidently not like that with cyber-writing. Here inscriptions seem readily reversible; traces can be destroyed and created with equanimity in the virtual spaces of cyberworld, for, as Derrida has intimated, the traces we work with are of presences that "never were."

The technology by means of which words and images appear on the computer screen is the same as in television. The images—instead of constituting a stable array of co-present elements—are projected onto the luminescent surface via an electron gun that scans the screen's pixels repeatedly; in this fashion, images are "refreshed" or reconstructed on an ongoing basis. In Chapter Two, I portrayed the medium of television as a prime exemplar of postmodernism, given its deconstructive/reconstructive, discontinuous, flattened, and ambiguous character. We are now able to see how the digital computer goes even further toward bringing postmodernism to its culmination. Although television involves the continual electronic "dematerialization" and "rematerialization" of visual images, the *sources* of these images are taken as decidedly "material." They consist in the "solid physical events" that are televised in studios or on-location. By contrast, computer images originate from the potentialities of cybernetic "flatland," Derrida's dimension of "horizontality" (1976, p. 69) in which all solid presence is leveled. It is the *virtuality* of the electronic images, their manifestation as traces of what never was ("footprints" not preceded by "feet"), that permits the astonishing flexibility of word processing. That is why the letters and words I type on this keyboard can be sculpted so fluidly; why I can add or subtract a word with such little effort; why I can move whole paragraphs or sections from one location to another, or merge these bodies of text—all in an essentially seamless manner, leaving no indications that any changes have ever been made.

Despite the advantages to be gained from engaging the virtual realm of the computer, the experience can be most disturbing, even more so than with television. In Chapter Two, I spoke of the addictive stupor in which I often find myself while viewing the "boob tube." Though I might like to believe that I watch TV just for "news" and "entertainment," it also implicitly serves as a mirror in which I seek self-definition. By covertly attaching myself to the images that appear on the screen (the characters, situations, issues, and causes), by identifying with them, I confirm my own identity. (The attachments in question are not necessarily positive. I define myself by my resentments, anger, hostility, and contempt, as well as by my positive preferences and affiliations.) But how vacuous are the offerings that the vacuum tube puts before me: an unending profusion of cartoon-like ambiguities; a dazzling variety of choices among nothing to choose, of differences that seem at bottom to make no difference. Since TV is my mirror, the underlying message is that *I* make no difference. Yet where else can I look but in the mirror, if I am looking to define myself? So, in watching TV, I engage compulsively in the attempt to fill myself from my own emptiness. This effort naturally fails, and, left entirely unsatisfied, I will seek satisfaction again and again in a viciously self-defeating cycle of addiction (see Chapter Two).

Nevertheless, with TV, I can take some comfort in believing that, behind the flat vacuity of the electronic tube, "real events" are indeed transpiring "out there" in studios and at remote locations. So while TV may be a "vast wasteland" (to echo the famous indictment issued by the media commentator Newton Minnow), a reassuring vestige of the old solidity is maintained. McLuhan did disclose that the "*medium* is the message," and the message conveyed by electronic leveling is one of de-realization. Yet, with TV, it is still possible for me to ignore this message and cling to the ideal of substantive presence—*my* presence, in particular. What I see in the "mirror" may be trivial. It may be fragmented to the point of being barely recognizable. Its profound ambiguity may well bespeak my own. And yet, if I am convinced that what I view on the monitor is "really out there," then I can at least conclude that *I* must really be "in *here*."

No longer is it possible to reach this conclusion in the starkly horizontal stretches of cyberspace. Because computer images undeniably *originate in absence*, in the equipotentialities of the virginal chip, the cyber-writer finds him- or herself on more perilous ground than does the viewer of TV. Indeed, in immersing herself in the utterly "weightless" field of *différance*, there really *is no* solid ground on which the writer can stand. Given that the virtual traces with which s/he plays are the phantoms of beings that never really existed, an unbearable lightness haunts the playing of this game. And the ephemerality of cyberspace can lead to an experience that surpasses the disconcerting anomie of television viewing: it can bring downright fear and trembling.

This is what I have discovered in my own exploits in the cybernetic abyss. My first full taste of it came in 1987 while working on my old Apple

IIe. I had invested myself in a conceptual project that had held great significance for me. Each day I had sat at my computer for hours on end wrestling with elusive subtleties, laboring to tease out daunting nuances of theoretical abstraction, and, in the process, taking great pains to shape just the right language to give coherence to my ideas. Then, one morning, with the errant press of a button, I overwrote the file that had contained my work and weeks of toil were simply obliterated. Not a catastrophic fire or an explosion, but the innocent stroke of a key had annihilated what I had previously put so much of myself into for so many days. This led me to radically doubt the weight of that work. The glaring discrepancy between my long sessions of effortful striving and the ease of their instantaneous effacement conveyed to me a sense of the uncanny. The feeling went beyond the ache of losing something valuable, and beyond the tangible fear that it could well happen again. My Herculean labors had been negated with such little difficulty that it led me to wonder how much real substance they could have had to begin with.

Given the colossal investment of myself in this work, its demolition symbolically simulated my own death. One does not survive such a "death." What has weight, substantive presence, *cannot* die. The laws of classical thermodynamics tell us that solid matter leaves its mark, and that this trace can neither wholly be expunged (the first law) nor returned to a state of potentiality existing prior to its actualization (the second law). But the cybertrace is evidently a ghost of what *never was alive in the first place*. There can be no "survival of death" here because there never was any palpable life. Ghosts of ghosts in infinite regress seem to be the only "inhabitants" of the cybernetic wasteland. In dwelling therein, I too become such a ghost.

Sad to say, my 1987 experience was not the last of its kind. Yes, I did contribute to that initial disaster by failing to properly back up my work. Since that time, I have been compulsively conscientious about saving to a floppy disk whatever I have written. But I have now "graduated" to more sophisticated technology and this has brought new perils. Today computers commonly contract "viruses," programs that are maliciously devised to disrupt data operations. An infection so incurred is not necessarily limited to the computer's hard drive, but can spread to floppy disk back-up systems as well. Not long ago, my computer caught a nasty virus (despite the anti-virus program that allegedly protected it). It was only by the narrowest of margins that I was able to avert the annihilation of months of work.

In general, the laptop computer I now operate behaves more erratically than did its "less powerful" predecessors. It flickers turbulently when booting up and shutting down, and sometimes fails to perform these tasks effectively. Inexplicable error messages intermittently materialize and programs become mysteriously "corrupted." I've told it all to countless support techs in more late-night S.O.S. calls than I care to remember. The "experts" frequently contradict each other, and, too often, they fail to provide any intelligible answers at all. Is my laptop a "lemon"? If so, the reports I hear through the "grapevine" tell me that the "lemon" has become a rather common "fruit" in today's computer markets.

Let me be clear that I do appreciate the power, agility, and communicative capability that my computer gives me. Yet, the whole time I operate it, there is a background sense of skating on thin ice. This feeling of insecurity has even invaded my sleep. On more than one occasion I have dreamed of working in front of the display screen when it suddenly goes black. This triggers an ominous sinking sensation in the pit of my stomach signaling that the "plug has been pulled," that the rug has been pulled out from under me, leaving me with no support. What remains is a phantom-limb feeling of active emptiness, gnawing absence. This is the downside of the cyberspace experience, its grave undercurrent. On the upside is that sense of levity, that heady feeling of exhilaration when at play beyond gravity's pull. The postmodern *modus vivendi* epitomized by cybernetic writing is appropriately captured by *Wiley Coyote*, a cartoon character familiar in popular culture. This comical creature dashes frenetically from place to place in hot pursuit of his avian nemesis *The Roadrunner*, until Wiley's precipitous, helter-skelter movements take him over a steep cliff. Defying the law of gravity, he continues his hyperactive doings in mid-air—until he notices that nothing supports him, whereupon he crashes to the ground with a thud. What an apt metaphor for the "game of the world" (Derrida, 1976, p. 50) played in cyberspace.

Now, according to Derrida, the emergence of cybernetics is not the only innovation that announces the liberation of writing from phonocentrism. Without elaborating, he alludes to "revolutions…in philosophy, in science, in literature" that have been taking place for "over a century" and that "can be interpreted as shocks that are gradually destroying the linear model" of writing (1976, p. 87). Of course, much of Part One of the present book is devoted to examining the twentieth century revolution in *physics*, and its implications for modern culture. The close parallel between quantum mechanics and cybernetics is brought out by Bohm (1986a), who develops at length the analogy between the information potential inherent in the computer chip and that which is contained in the quantum mechanical wave function. As with the silicon chip, "the wave function contains information implying a vast range of potential or virtual activities. In this case, these will be actualized by entering into the energy of the self-active particles" (1986a, p. 126). It would be something of an oversimplification, however, to equate Bohm's concept of information potential with that of Derrida.

Bohm's approach to physics is influenced more by Einstein than by Bohr. While Bohm rejects the old mechanistic doctrine of causal determinism, he takes to heart Einstein's declaration that "God does not play dice." The compromise to which this leads is an interpretation of quantum mechanics in which causal forces are still at play in the universe, though these forces are less obvious than the gross mechanical ones of Newtonian classicism.

Bohm initially expressed this view in his "causal interpretation of quantum mechanics" (1952). Whereas the thinking of the Copenhagen School leads to the conclusion that sheer uncertainty lies at the core of the wave function (see Chapter One), for Bohm, quantum functioning entails a subtle form of certainty. It is not outright disorder but "hidden variables" that underlie the pushes and pulls of explicitly measurable bits of matter. This formulation evolved into Bohm's (1980) well known concept of "implicate" and "explicate order." The latter corresponds to the manifest Newtonian world wherein isolated bodies appear to interact with each other through forces that are strictly external. What quantum mechanics shows us, according to Bohm, is that the explicate order of gross matter-energy actually originates from, is enfolded within, a subtler, implicate order that is mind-like in nature. The implicate order of Bohm's "causal interpretation" entails neither acausal formless chaos nor mechanically determined external causation, but the *self*-causation of *form*, of *in-formation.* Here informational potentials are self-actualized in an essentially Aristotelian fashion. Bohm's *potentia* pre-exist more solidly than Derrida's ephemeral traces; in a subtle sense, the *potentia* are already present within the implicate order; they are "in there" waiting to unfold themselves. We recognize the self-causal agency of Bohm's implicate order as paradigmatic of *modernism.* For its part, the Copenhagen School can also be viewed as modernist, but in a more marginal way; instead of denying chaos outright, it adopts a "pragmatic" strategy of "working around" the chaos (as discussed in Chapter One).

In the present chapter, we have seen how modernism leads inescapably to postmodernism, since the self-objectification involved in the former implicitly brings a re-subjectification that opens the door to the infinite regress. We see this in Bohm as well. In making the transition from the Newtonian order of explicate matter to the implicate domain, Bohm has no intention of simply relinquishing objectivity, but wishes instead to develop new forms of mathematical expression that will describe the mind-like interior realm in a completely objective manner. The re-subjectification that follows is evident in the need Bohm feels to "extend and generalize...the implicate order" (1986b, p. 196). He discloses that there is not just one implicate order serving as "the ultimate and unique ground of all reality," but that "There will... have to be yet deeper and more comprehensive orders within which this context is embedded" (p. 196). Furthermore, "Because there is no 'bottom level,' any given level is not *completely* grounded in any other specifiable set of levels, nor is its explicate aspect...grounded completely in any particular implicate order" (p. 198). As a consequence, "Each context has a certain ambiguity....Everything is ultimately open, as new contexts may always come in" (p. 198). In short, every objectification that explicates what had been implicate or subjective brings with it a new order of subjectivity—*ad infinitum*. Thus, while Bohm apparently had every intention at the outset of carrying the banner of Einsteinian objectivism, in the end he succumbs to a Derridean subjectivism.

Bohm does attempt to extricate himself from the relativistic labyrinth. He asks us to consider that, beyond the open hierarchy of "implicate-explicate levels," lies the closure of an "infinite totality" that consists of "the whole movement (holomovement), which includes all contexts and their relationship...[and] is creatively self-determined" (1986b, p. 200). Acknowledging that the effort to describe this "holomovement" may bring us to "what one may feel to be a paradox," Bohm advises us to seek "another approach that does not lead us into this paradoxical area" (p. 200). We get tangled up in paradox because we try to gain closure on transcendent totality by relying on conceptual description. In this approach, every postulation of totality brings its denial, since our dependence on symbolic abstraction confines us to the open-ended mirror play of explicate and implicate orders. What is required instead of abstract language is an act of *concrete perception*: We must attempt to perceive "what may be reflected in...the implicate-explicate levels....Each kind of order...acts like a mirror, but the two kinds together point more accurately and more deeply into the infinite totality than either alone could do. From this perception consciousness may turn toward the infinite inward depths that are its ultimate ground and, in some way, come into actual contact with these depths" (p. 200). So Bohm is indeed still looking for an "ultimate ground." He is hoping to find his way out of the hall of mirrors, and—through the self-intuition of a wholly transparent consciousness—arrive at unmediated presence.

But *can* we simply leave the relativistic province of the intellect, cross over its boundary, and immerse ourselves in the absolute, in the perceptual purity of the infinite? Do not such boundary crossings themselves create *finitude*? Are they not acts of simple division that express the divisiveness of the *intellect*? What Bohm tries to do is *abstract* himself from the impurities and paradoxes of the abstract. In attempting to exit the regress of mirrorings in this way, he remains squarely within it. It is like the proverbial Chinese finger puzzle: the more we struggle to free ourselves from paradox—from that "impure" serpentine creature, that hybrid dweller of the labyrinth whose name is *apeiron*—the more we are entrapped in it. The very "purity" of Bohm's struggle for concreteness betrays its abstractness. Bohm himself has stated the rule of the mirrored labyrinth: every one-sided positing of presence brings with it an absence, every such closure an opening. And he proceeds to operate in *accordance* with this rule when—in his desire to avoid the *paradox* of closure and openness—he posits one-sidedly an ultimate *closure*, an "infinite totality" that is the "ultimate ground." How could an affirmation that adheres in this way to the law of the regress bring anything but its own negation? The crux of Bohm's problem then, is that his very manner of positing a departure from the hall of mirrors guarantees his entrapment there.

Derrida's inhabitance of the regress also is guaranteed, not by one-sided affirmation but by its mirror counterpart: a one-sided negation. Like Bohm, Derrida evidently wishes to avoid the "impurity" of paradox. Therefore he asserts that we should not call the "presence-absence of the trace…its ambiguity but rather its play (for the word 'ambiguity' requires the logic of presence, even when it begins to disobey that logic)" (1976, p. 71). Apparently, the logical hybridity of a term that undermines its own requirement is abhorrent to Derrida. It is *purity* that Derrida is seeking, a pure negativity or absence unadulterated by being paradoxically blended with presence. Although this shift to the negative does not free us from metaphysics, as Derrida has claimed, it does actually bring us a step closer to freedom, as we will soon see. But let me first spell out the sense in which Derrida, and postmodernism in general, remain *within* metaphysics.

There is little doubt that postmodernism reverses metaphysics, turns all its affirmations into negations. But what seems more essential to postmodernism than its choice of the negative is the categorical manner in which it makes this selection. In choosing unambiguously, in emphatically coming down on one side of the binary pair as opposed to the other, postmodernism preserves the underlying one-sidedness of metaphysics, its drive for logical purity. Thus logocentrism appears to be maintained through all the reversals and overturnings carried out by "deconstruction," Derrida's statements to the contrary notwithstanding.

The world of postmodern philosophy has been set abuzz by Derrida's recent "conversion" to a kind of "negative theology" reminiscent of the Middle Ages (see Caputo, 1997). Is there not after all a compelling similarity between the absolute elusiveness of God proclaimed by medieval mystics like Meister Eckhart, on the one hand, and the endless deferral of presence characteristic of Derridean *différance*, on the other? Whereas the early Derrida (1967/1982) had completely dismissed such parallels as superficial, he has now (1995) apparently relented. If Derrida has in fact "converted" from the total negativity of secular deconstructionism to the implicit positivity of "negative" theology, this switching of sides in itself does nothing to challenge the inherent one-sidedness of metaphysics. In choosing as he does, whichever side Derrida selects becomes more basic than its opposite; it becomes his basis, his ground, and a ground is a ground, even if it is deemed "groundless." As Nietzsche understood, there is really no separating metaphysics from religion, the most essential characteristic of the latter being its obsession with purity (Felser, 1999). What we are seeing then is that, even with the apparent radical secularity and relativism of the postmodern, God the Father is still not dead. Owing to her "impurity," the goddess *apeiron* is still denied.

Postmodernism's penchant for one-sidedness is undoubtedly related to its abstract stance. By attaching myself unequivocally to one member of a binary pair, a part of me is *de*tached. In choosing one—be it presence or absence—I make it *present*, and, in the process, *I* remain absent. In choosing what had been implicit, I make it explicit, *objectify* it, whether I call it "objective truth" (as in modernism) or "subjective play" (as in postmodernism). This is an act that surely brings with it new implications, a new subjectivity at a higher level of abstraction. Derrida does not merely choose differences that are then supplanted by identities in endless succession; he chooses difference as such, *différance*. In being so chosen, this "higher-order difference" is made into a "higher-order identity," a "hyper-object" cast before the gaze of a new subject: the *Derridean cogito*, who might indeed be likened to the God of "negative theology"—He for whom the "hyper-absence" is present. So goes the "game of the world." In this interminable play, the old formulation of subject and object is recycled in unlimited subfractions of itself, like the epicycles of Ptolemy. But never is the formula surpassed.

And yet there is a sense in which the postmodern choice of the negative side, while not surmounting the one-sidedness of metaphysics, does tend to subvert it. We have seen that modernism never really accomplished what it had intended. What began in the middle of the nineteenth century as an attempt to reduce ambiguity by abstractly maintaining the positive (determinate law, fixed identity, scientific objectivity, etc.), ended in the middle of the twentieth century with a lapse back into ambiguity. This is where postmodernism enters the picture. Postmodernism is also an attempt to reduce ambiguity. But it must deal with the fact that the loss of solid presence occasioned by the collapse of modernism is irreversible. Unable to eliminate ambiguity by simply reverting to the positive, postmodernism must attempt to achieve this by one-sidedly affirming the negative: subjectivity, difference, "freeplay" (Derrida cited by Spivak; 1976, p. xix). Why does this *undermine* the one-sidedness of metaphysics? It is essentially because the tenuousness of one-sided choice—which was certainly implicit in modernism from the start—now becomes patently obvious.

Although it can be said that modernism and postmodernism alike act one-sidedly to turn the subject into an object, postmodernism cannot accomplish this simply by going back to the *old* objectivity. Instead, in its *mirror reversal* of modernism, postmodernism turns the subject into a *virtual* object, an "*anti*-object." Like the anti-matter of which physicists speak, the postmodern "anti-object" is unstable; it is given to being annihilated suddenly, in a way that transmutes it into its opposite. As a consequence, the ambiguous oscillation between objective and subjective poles that was only implicit in modernism now becomes unmistakable—even though Derrida, in his drive for purity, might like to continue denying it, as when he one-sidedly emphasizes the *subjectivity* of the trace (its negativity, its absence), rather than its ambiguity per se.

Despite the fact that Derrida categorically chooses the negative as Husserl had chosen the positive, the postmodernist lacks his predecessor's conviction. There is a note of ironic humor in Derrida that is conspicuously absent from Husserl's more straightforwardly earnest approach. Yet it seems the joke is as much on Derrida himself as on his readers. Or we may say that he is involved in what Jean-Paul Sartre termed "self-deception" (1943/1975, p. 299ff). That is, as Derrida focuses on the negative, he is aware of the positive side, but only peripherally, in a way that allows him to dismiss it. If the positive were completely opaque or wholly transparent to him, there would be no self-deception; in the former case, he simply could not lie about the positive, whereas, in the latter, he could at least not lie to *himself*. The game of self-deception depends on the "translucency" (Sartre, p. 302) of the member of the dyad upon which one is not focusing. At the very moment Derrida is proclaiming that "all is absence," the instability of this absence is such that the light of presence is diffusely filtering through to him. But—in order to banish ambiguity and sustain one-sidedness—he blinds himself to this light. I am not suggesting that Husserl and all who preceded him were not also in self-deception. The point is that, since the collapse of modernism, it has become more difficult to banish ambiguity convincingly, so that the self-deceptiveness of choosing one-sidedly has become more obvious.

What effect does it have on me when—at a level not too far from the surface of awareness, I *know* that I am deceiving myself, that my choices are virtual? Does this prompt me to drop my one-sidedness and accept the ambiguity essentially at play? Ambiguity is the last thing in the world I want to accept. Therefore, if I can no longer convince myself of my solid presence, I would much sooner acknowledge my ghost-like insubstantiality than admit to an ambiguity at my core. For, it is not so much absence, "pure negativity," that "I"—this *ego cogito*—find so terrifying, as it is the "impurity" of *apeiron*. Since I just cannot bring myself to accept *apeiron*, there seems nothing else for me to do but continue my attempt at self-deception, despite its glaring transparency now.

It is this self-deception that underlies the ravages of addiction that have swept our postmodern world. Addictive substances lack solidity. They are *virtual* substances, "*anti*-substances" (as in the "anti-*heroes*" that populate postmodern fiction). It is *because* of their ephemerality that we get "high" on them. The more ambiguous the substance, the "higher" we become. For, while it is indeed more difficult to deceive ourselves by holding to one side under conditions of greater ambiguity, when we *can* succeed at doing this the relief is that much sweeter. An increase in ambiguity can be said to have the effect of increasing the frequency of fluctuation between opposing sides. When the oscillation rate is low and we can easily fix our attention on but a single side, although there is surely less ambiguity, there is also a lesser degree of contrast between sides, since the other side is essentially out of the picture. As the shift from side to side becomes faster, sides are brought into a common frame of reference; they come to be juxtaposed in such a way that the contrast between them is brought out in stark relief, giving each side a sharper definition and simplicity, when taken by itself. Therefore, on those moments when the ambiguity is resolved in favor of one side, the apparent clarity we achieve is dramatically enhanced. Does the particular side that is favored by the resolution have any effect on whether or not the addictive "high" is maintained? Evidently not. Whether the resolution is "positive" or "negative," what matters is the apparent clarity that is brought in dispelling the ambiguity, at least for the moment.

A good example of this is the form of addiction that involves taking risks. The gambler keeps coming back to the casino because of the intoxicating high s/he gets in a situation where, again and again, s/he knows the thrill of passing from extreme ambiguity to one-sided clarity—even when the clarity is gained by losing. For, it is not winning or losing per se, but the *act of resolving ambiguity* upon which the addiction is based. Therefore, the greater the ambiguity that is created, the more satisfying is its resolution. It seems that electronic addiction works in essentially the same way. The high I get from watching TV or operating this computer depends on the inherent ambiguity of these media, as discussed above. I obtain higher resolution, a higher, more vivid definition of what is experienced (as in high-definition TV), by virtue of the fact that the underlying digital micro-technology—involving as it does a closer approximation to the Planck length—entails greater uncertainty. From this we can better understand why quantum mechanics has been so successful (see Chapter One). The unprecedented precision of microphysical research derives from the fact that it resolves unprecedented levels of uncertainty. In short, the more chaotic the system, the more markedly discontinuous the quantum leaping, the more order and continuity appear to come out of it when the ambiguity is resolved.

What we are now able to see is that postmodernism's addictive high does not come simply from avoiding the irresolvable ambiguity of concrete *apeiron*, but from simulating that ambiguity abstractly, in a *resolvable* form. With every character I type on this keyboard, the underlying ambiguity of the microprocessor is resolved, and, because that ambiguity is so high, this reduction of ambiguity I am engaged in elevates my mood. Nevertheless, at the same time that heightening ambiguity brings heightened levels of reward when the ambiguity is resolved, it also brings me closer to the point where the ambiguity *cannot* be resolved by my one-sided self-deceptions. That is why these electronic feats of levitation that I perform have such a tenuous feel to them. It is why, like Wiley Coyote, I dare not look down, lest the screen go black, the computer crash—as it does tend to do. In this way, we go from "booting up" to "crashing"—or "*shooting* up" to "crashing," in the case of the drug addict and the stock market. In Chapter One, were we not dealing with the very same transition from resolvable to irresolvable ambiguity, from drunken levity to sober gravity, when we examined the history of modern physics? Proceeding ever more deeply into the microworld, higher and higher degrees of uncertainty have been encountered in physics, along with ever more spectacular precision—until the Planck length has been reached, whereupon the equations have gone infinite, precision has been utterly lost, the ambiguity no longer being resolvable. Since QM's *irreducible* ambiguity is associated with problematic *gravitational* effects (quantum gravity), we may well say that this high-flying scientific discipline has now crashed into a black hole. And yet, physicists are still attempting their feats of levitation (as discussed in Chapter One). It is true in general that, while the same addictive substances that make us high assure our fall, we keep coming back for more. What else can we do in the virtual world of postmodernity? Though we get drunk on every one-sided choice we make, soar to the giddy heights, only to crash back down again, is there any other viable option short of what seems wholly unthinkable to us: to truly and fully accept the ambiguity of our existence?

This ambiguity doubtlessly betokens *apeiron*. She draws closer to us now than ever before and we tremble in her ominous presence, going to any extreme to avert her Medusan gaze. Alongside the addictions of science, technology, and popular culture, a similar one-sidedness predominates in contemporary philosophical thought, as we have amply seen. Beyond Derrida, who has been our exemplar, we find the inverse intellectual puritanism variously operative in the cultural deconstructions of Jean-François Lyotard and Gilles Deleuze, in Julia Kristeva's subversion of patriarchal psychoanalysis, in the hyper-skeptical philosophy of Richard Rorty, and in the "language games" of Ludwig Wittgenstein, to name a scant few. Postmodernity's "game of the world" is surely a *head* game, excluding as it does *apeiron's* dark body. One imagines Nietzsche's Zarathustra becoming increasingly *light-headed* as he giddily climbs to higher and higher peaks. The higher he goes, however, the more he feels the pull of his "spirit of gravity," the "homunculus" that weighs him down, the serpent that nauseates him and fills him with dread. For snakes are "'land animals with no legs which lay eggs'" (Leach quoted in Berman, 1989, p. 80). Thus they "fall into the category of the ambiguous or the anomalous….Such animals, says Paul Shepard, 'offend the necessity of clear distinction'" (Berman, p. 80). We fear the snake, notes social historian Morris Berman, because it is "disturbingly… 'intermediate'" (p. 77); it "transgresses categories" (p. 80). To be sure, the serpent is *apeiron*, and what she essentially violates is the purity of Zarathustra's Faustian mind-quest. Zarathustra evidently still believes that "cleanliness is next to Godliness," and, if *apeiron* is anything, she is "unclean."

The price Nietzsche paid to avoid the Black Madonna was division unto madness. It is the price we are still paying in this schizophrenic postmodern world. Like the shepherd in Nietzsche's "Vision and Riddle," we have been unable to swallow the uroboric serpent and thus have "bitten off its head" and "spewed it far away." It is that disembodied head that now rules us in the head games of our lives. We are addicted to the light; to the clear and the clean; to categorial purity—"addicted to perfection," as Marion Woodman (1982) puts it in her Jungian reading of the contemporary epidemic of anorexia that shrivels the human flesh (particularly the flesh of women). The greater the ambiguity, the more desperate we are to reduce it, to the point where we are willing to pay for "clean-cut" value and meaning by cutting ourselves down to the bone.

My own addiction is obvious to me in the conceptual game that I have been playing for many years now. It is the game that governs *the writing of this book*. All too often I deceive myself, seeking perfection one-sidedly, striving for the purity of simple closure. It is what I am tending toward *now*, as I draft this text. In searching for the perfect words and phrases to turn this thinking process into a cleanly finished product, I am frequently so fearful of the messiness and murkiness of *apeiron* that there are moments when I can barely bring myself to write anything at all. It is hardly surprising, given the fundamental ambiguity of my subject matter, and indeed, the ambiguity of this cybernetic medium in which I work. What a "high" I get when it appears I am on the verge of bringing order to this chaos! And how often I have fallen from the heights of "categorial clarity" into the yawning abyss. It is then that I feel the ache of an active emptiness. Continuing in free fall, I feel totally exposed and vulnerable, wholly without support. There is often a mood of sheer desperation that accompanies my ongoing crash into groundlessness, an urgently felt compulsion to find something—anything!—to hang on to. When this happens, every instinct tells me to break the fall, to reach for some overhanging ledge and cling to it for dear life. But would that not amount to still another effort at denying *apeiron*? It seems then that some existential "sky-diving" is in order. In aiming to surpass modernism and postmodernity alike, we must stay in "free fall," however perilous that may appear to a panicky *cogito* like myself. The "pull of gravity" surely is strong. It is an impulse that no ground can stop, since what draws us in is the *black hole*. Yet, despite our grave trepidations, it seems we must find a way to accept the dark chaos of *apeiron*.

Before considering the question of just what such acceptance would entail, let us attempt to refine our understanding of what we fear. To be sure, it is the ambiguity of *apeiron* that terrifies us. However, in previous pages *apeiron* has been associated both with ambiguity (uncertainty) and with *contradiction*. To begin clarifying the relationship between these terms, recall Kierkegaard's distinction. On the one hand, there is "uncertainty," and, on the other, we have "absurdity" (Kierkegaard, 1846/1975, p. 117), meaning contradiction. In unpacking Kierkegaard's analysis of these alternatives, I noted in Section 1 of this chapter that, while uncertainty raises doubts about objectivist truth, it still leaves open the possibility that those doubts might be dispelled. With contradiction however, there is no longer any doubt about the inadequacy of objectivism. Restated in Sartrean terms, we might say that it is when we are confronted with outright contradiction that we can no longer effectively deceive ourselves. For the deception to be perpetrated, the opposing perspective cannot be completely transparent but must be "translucent." That is, I am able momentarily to convince myself of my one-sided "truth" only when the other side is relegated to the background, when it is given to me in but a peripheral way. Because the other side does register in such a case, there is that feeling of ambiguity. Nevertheless, I am still able to hold my doubt at bay and "cling to a ledge," at least for the moment. No longer is this possible when the other side becomes totally transparent, when it enters the foreground in such a way that opposing sides are blatantly juxtaposed upon one another. We might say that, at this point, the ambiguity becomes "infinite," giving way to the uncanny sense of contradiction. If the postmodern increase in ambiguity is thought of as increasing the frequency of fluctuation between opposing sides, as I suggested above, then the transition from ambiguity to contradiction may be seen as a passage to an *infinite* fluctuation frequency—as in fact happens in microphysics, when passing the Planckian limit of manageable uncertainty introduces infinite fluctuations into the field equations that render them unintelligible. This is the point at which ambiguity is utterly irreducible by the abstract intellect. It is the place where the heady *cogito* inescapably confronts the bodily *apeiron*. It is the place the *cogito* fears most. And, as Kierkegaard understood (though in a manner constrained by his Christianity), this is the place I must fully inhabit, must make my home.

To the *cogito*, however, *apeiron* surely has the look of the Medusa. When he gazes upon her from his abstract perspective, he witnesses the horror of horrors: a head full of tangled snakes. This symbolizes the ruination of categorial reason via endless ambiguity, the absolute impurity of unbridled contradiction. From the posture still seeking to *objectify apeiron*—to set her up as what is seen in her totality by a subject that is not seen at all, as what is simply present to him who is simply absent—the uroboric goddess surely cannot be stomached. Any attempt to accept *apeiron* while maintaining this phallic stance can lead only to her continued denial. Derrida is a case in point. We have seen how his ultimate refusal of ambiguity turns his purported embrace of presence-absence into a one-sided absence (*différance*). In seeking to accept *apeiron* on *his* terms, he fails in the end to accept her at all —which causes him to slip back once more into her Medusan grasp, since his "pure negativity" is all too susceptible to flipping over into "pure positivity" (recall his "conversion"). To accept *apeiron* in earnest, she must be accepted on *her* terms.

Up to this point in the text, I have done little more than give vague intimations of what such acceptance would mean. Yes, we must attune ourselves to the maternal rhythms of nature, heed the "spirit of gravity," surpass the puritanical striving for one-sided unity, and hold the contradiction of unity-in-diversity. But how, exactly, are we to enact these gestures? If they *could* be carried out, the very same contradiction that currently devastates the *cogito* might metamorphose into a paradox that is fruitful and restorative. For we have seen that the serpent or *lapis philosophorum* is at once the "lowest and the highest." The "arcane substance" is primitive and undeveloped—"prima materia," the "primeval chaos" (Jung, 1967, p. 235)—and at the same time is "the panacea, the universal medicine…the tincture that transmutes base metals into gold and gravel into precious stones" (Jung, 1967, p. 101). Just what is it that we must do then, to transform the vicious circle of pandemic addiction into an uroboric cycle possessing generative powers? How can we go from the Medusan serpent to the Black Madonna, from "Bad Mother" to "Great Mother"?

Tricky are the ways of the serpentine *apeiron*. Indeed, she is equivalent to the Trickster of Greek mythology (see Sherman, 1993). In alchemy, *apeiron* takes the form of the Spirit Mercurius. The Mercurial challenge is epitomized in the story of the "Spirit in the Bottle," the alchemical tale of individuation recounted at the end of the previous chapter: The tempestuous Spirit has escaped from a strange bottle and menaces the boy who mistakenly had liberated it. Somehow, the boy must trick this Trickster back into the bottle. But how is this to be done when the Spirit is *so* mercurial that every effort to close her in only results in her slipping away? The alchemist Nietzsche could not quite solve this riddle, given his penchant for scaling the heights, his one-sided preference for levity over gravity. Husserl fared no better at containing Mercurius, nor did Derrida. The lesson evidently is that no effort at puritanical closure—be it positive or negative—can confine the protean deity; from every attempt at simple containment she slithers away.

But the alchemical literature hints at a *different kind* of containment vessel, one whose contents are hermetically enclosed by *allowing* them to spill out! Jung introduced his discussion of the Hermetic vessel by noting that, "Although an instrument…it is no mere piece of apparatus. For the alchemists the vessel is something truly marvelous: a *vas mirabile*….[T]he whole secret [of alchemy] lies in knowing about the Hermetic vessel" (1968, p. 236). In my detailed study of the vessel (Rosen, 1995), I established that its structure is *paradoxical*, and, more specifically, that it possesses the structure of the *uroboros*, the snake that swallows its own tail and that, in so doing, *contains itself* (Fig. 5). Jung made it clear that this curious Hermetic container is none other than the bottle in the alchemical fairy tale (see Chapter Two). What is the relevance of all this to the challenge of accepting *apeiron*?

In his effort to contain *apeiron*, what the puritanical *cogito* cannot swallow or keep down, is the fact that *apeiron contains herself.* Thus, when he fails to tame her, to keep her "bottled up" one-sidedly, he violently expels her. For, if she is not his "good little girl" or "Good Mother," she is a witch, a Medusa, a demon that must be cast out. In this way, by continuing to insist categorically that *apeiron* is either inside of him or outside, either simply contained by him, or simply on the loose (a "loose woman"), the *cogito* goes on denying the paradoxical truth of her *self*-containment. And this is just what he must find a way to accept. But how, exactly, is he to do it?

Four

APEIRON AND BEING

[I]n order to bring about their subsequent reunion, the mind (mens) *must be separated from the body...for only separated things can unite. (Sixteenth century alchemist Gerard Dorn, as paraphrased by Jung, 1970, p. 471)*

1. The Ontological Difference

Coagula cannot properly be thought when taken in categorical opposition to *solve*, since, to do so would only attest to the continuance of *solve* alone. Turning to embodied *apeiron* cannot simply mean turning away from the abstract *cogito*, since any such simple turning away is itself the mark of abstraction, as we have seen. In fact, it is not just that abstraction cannot be abandoned in engaging *apeiron*. To enter *apeiron* in earnest, abstraction must be carried to its *fullest expression*. It is for this reason that the most abstract reflections of that most abstract discipline, namely, philosophy, are what wind up leading the way to concrete *apeiron*.

By and large, the historical trajectory of post-Enlightenment philosophy traced in the previous chapter mirrors what was outlined in Part One: *Apeiron* resurfaces at mid-nineteenth century, there is an initial flirtation with her, quickly followed by an attempt to repress her (modernism) that proves futile in the end (postmodernity). What we are now prepared to consider is that—between the advent of Husserlian modernism and its collapse in Derridean postmodernism—another alternative is foreshadowed.

Husserl's most promising student was the existential phenomenologist Martin Heidegger (1889–1976). In the essay "My Way into Phenomenology" (1963/1975), Heidegger acknowledges his teacher's influence, then contrasts himself with Husserl in no uncertain terms: "Whence and how is it determined what must be experienced as 'the things themselves' in accordance with the principle of phenomenology? Is it consciousness and its objectivity or is it the Being of beings...?" (1963/1975, p. 239). Heidegger had been drawn to Husserl in the first place because he had seen in phenomenology the possibility of addressing this question of Being that was his all-consuming passion. What he found, to his disappointment, was that—even though "Husserl's thought was determined by Franz Brentano," and Brentano's dissertation on Being had been Heidegger's "chief help and guide" in his "first awkward attempts to penetrate into philosophy"

(1963/1975, p. 234)—Husserl himself had no real appetite for tackling the matter of Being. Lauer characterizes Husserl's neglect of Being as "a strange paradox never resolved by Husserl and that he seems never to have tried to resolve" (1965, p. 46). Although Husserl clearly saw the importance of the question of essential truth for his philosophy, "he never really asked himself the question" (p. 46). Why not? It is because "his philosophy is primarily concerned with the essence of what is (*das Seiende*) rather than with the essence of being (*das Sein*)....His problem, then, is not to discover the essence of either being or truth [*Wahrheit*], but rather to guarantee that knowledge is of 'what is' and that it is true [*wahr*]" (p. 46). In other words, because of his rationalist preoccupation with obtaining "absolute scientific certainty," Husserl was unconcerned with being *qua* being; his abiding interest was in being *qua* apodictic knowing. To Heidegger, on the other hand, the question of the "Being of beings" is philosophy's foremost concern.

Heidegger's thinking takes its rise from the *ontological difference* (1927/1962). The crucial distinction is that between the "ontical" and the "ontological." Although Heidegger himself provided no explicit definitions of these terms, his translators did: "Ontological inquiry is concerned primarily with *Being*; ontical inquiry is concerned primarily with *entities* and the facts about them" (Macquarrie and Robinson in Heidegger, 1927/1962, p. 31, n.3). According to phenomenological philosopher David Michael Levin (1985), while this "fundamental difference...between Being and beings...is both basic and simple, its articulation and understanding are matters of the greatest difficulty" (p. 10). Levin proposes:

> [A] fruitful way of formulating the ontological difference is to articulate it as the difference between the horizon, field, or clearing within which beings appear, and the various beings themselves; or as the difference between the ground of significance itself and the figures which appear in its setting and stand out from the ground. Being is not *a* being, but rather the *dimensionality* within which all beings are to be encountered. (p. 11; italics added)

The difference can be clarified further by recognizing that, if Being is a dimensional context or background, it cannot be so in the same sense that classical space serves as background. For Being is not just the ground from which *figures* stand out; it is not merely that which functions as a framework or container of objects; rather, the objects *and* their spatial background emerge from Being, along with the detached subject that reflects upon those objects. In other words, what stands out from Being, what Being opens up and first makes possible, is *object-in-space-before-subject*.

Seen in this light, the "universal essences" of Husserl's "transcendental phenomenology"—which appear as objective structures in a space of "pure logic" cast before an abstractly self-reflective subject—are merely ontical. *Heideggerian* phenomenology requires the *ontological*: the awareness of the pre-reflective Being that first gives rise to object-in-space-before-subject. The difference between these two approaches does not just concern the philosophical content of thinking. At stake here is something deeper, something that speaks to the underlying manner in which thinking operates: its posture, orientation, or gearing. In classicism, modernism, and postmodernity alike, thinking is geared to move "forward." It is poised to move *from* prereflective Being *to* the reflective structure constituted by the old subject-object opposition. In this orientation, the *cogito*, while starting out from Being, is fundamentally inclined to move away from it. As a result, Being per se is eclipsed. The movement away from Being is similar to the "from-to" action of experience described by the phenomenologist Drew Leder (1990, pp. 15–17). Ordinarily, in whatever we see, we "cannot see our own seeing" (p. 17), since our seeing is what we see *from*. By upholding the categorical distinction between *what* we see and what we see *from*, the subject-object dichotomy is enforced: sight is directed *to* an object seen, *from* a subjective perspective. The movement is assumed to be irreversible, making it impossible to view subjectivity as such (note the connection with the Michelson-Morley experiment, as discussed in Chapter One). Is Heidegger suggesting that we challenge this assumption of irreversibility by transposing the subject-object relation in such a way that we can know the *subject* per se? It was the modernist Husserl who had hoped to know the subject as the object had once been known. For his part, Heidegger does not merely call for a reversal of the inclination to pass from subject to object; he calls for a reversal of the *subject-object dichotomy*, thereby enabling us to move back into *Being*.

The *gearing* of awareness is all-important to the reversal in question. In order to apprehend Being, it is not enough to change the *direction* of thinking; its "gear" must be reversed so that it is no longer simply moving ahead. When proceeding toward Being in the "forward gear," we attempt to "turn around" upon this prereflective source—like turning a car around so that it now faces the direction *from* which it was previously facing. In such a reversal, we are indeed inclined to *face* Being, to have it appear before us, over against our consciousness. With our thinking thus geared, we still try to *objectify* Being, but only succeed in obscuring it. We cannot know Being by simply turning around upon it any more than turning a car 180° to face the direction *from* which it *had* been facing allows us to capture that "from" as it was initially experienced; such a turning merely turns the old "from" into a new "to" that is seen *from* a new perspective. It is clear that the attempt to grasp the "from" of Being while maintaining the forward orientation is futile; to make such an effort is to "chase one's tail," to turn in a vicious circle. The implication of Heidegger's work is that thinking must shift into a different gear; if we are to apprehend Being *qua* Being, we must approach it in "reverse." Notice that, whereas turning a car around to face in the opposite direction turns us away from the original direction in which we faced, if we shift into reverse, we continue to face in the same direction. A similar distinction can be made with respect to Being: In seeking to come "face-to-face" with it, we hope (in vain) to turn our backs on the subject-object dichotomy, whereas the movement *backward* into Being does not simply negate the reflective formula but brings to light the prereflective ground from which it issues. Before too long, I will go into more detail on the precise nature of "thinking in reverse."

In the forward orientation, Being contracts to a "nullpoint" (Leder, 1990, p. 13), a mere point of departure for object-in-space-before-subject. The latter then can come to be taken as *self*-initiating, as that which is always already given. It is by moving out of and away from the prereflective ground that Husserl can set his sights on a transcendental consciousness that is universal, necessary, and perfectly objective (he does admit, however, that it might take centuries of phenomenological research to approach this pristine state of consciousness!). When Husserl thus indulges himself in the Faustian vision of total self-sufficiency and self-consistency, reality is idealized as a structure of pure logic, a frozen world of apodictic certainty. But what if the orientation of thinking could be reversed? asks Heidegger. What if the ontical gearing of thought could become ontological? Then what is otherwise a mere point of departure for a static world would become a whole dimension of vital activity unto itself. It is clear that, for Heidegger, Being is not simply a subject, an object, or a spatial continuum. Instead, it involves the domain of prereflective action that first makes possible these seemingly fixed terms of reflection. Perhaps then, Heideggerian Being might be written as "Be-*ing*," to indicate that it is essentially a dimension of *dynamic process*. But, if that is the case, why not simply call it "*Becoming*"? The answer lies in the paradoxical character of what Heidegger designates "Being."

For Heidegger, the movement backward into Being is traced by a backward movement in *history*. Because the history of the West is at bottom a chronicle of Being's progressive concealment, only in the earliest moments of this epoch could the echo of Being still have been heard with any clarity. Accordingly, Heidegger brings us back to what "is considered the oldest fragment of Western thinking" (1946/1984, p. 13), the *Anaximander fragment*: "'But that from which things arise also gives rise to their passing away, according to what is necessary....'" (p. 20). Heidegger's meditation upon this twenty-five-hundred-year-old passage leads him to focus on the *Being* of the "that from": the things or beings arise, become present, from

Being. However, while the beings surely are not simply the causes of themselves, they are not brought to presence by the action of some external agency. This is implicit in an inherent ambiguity or confusion in the Greek terms, *όν* and *ειναι*, "being" and "to be." Heidegger tells us that, if we are to open a genuine "conversation with early thinking," we must remain "within the abiding confusion," rather than "drifting right on by the confused state of affairs," as all "Being-talk" ("[b]uoyed by the formal correctness of the translation of *όν* and *ειναι* by 'being' and 'to be'") has tended to do since those early days (1946/1984, p. 25). Apparently, the term *όν* does not refer to "being" in a manner that sharply differentiates it from "to be":

> *όν* says "being" in the sense of *to be* a being; at the same time it names a *being* which is. In the duality of the participial significance of *όν* the distinction between 'to be' and 'a being' lies concealed. What is here set forth, which at first may be taken as grammatical hair-splitting, is in truth the riddle of Being. (1946/1984, pp. 32–33)

A little later Heidegger declares that "since the dawn of thinking 'Being' names the presencing of what is present" (p. 39). Putting these two passages together, the "being which is" is that which simply is present, whereas "'being' in the sense of *to be* a being" describes the process of presencing. Given the overlapping within the single term *όν* of the two senses of being, Heidegger's assertion that "'Being' names the presencing of what is present" must be read paradoxically. This can be facilitated by adding parentheses to the statement: "'Being' names (the presencing of) what is present." Excluding what appears within parentheses gives us the *reflective* aspect of Being: in its capacity as what simply is present, Being constitutes the beings, things, or objects arrayed in space before the reflecting subject. In the mode of "what is present," then, Being functions as object-in-space-before-subject. When we *include* the parenthetic phrase, the *prereflective* aspect of Being is given: it is the *process that makes present* what is present, the act that first gives rise to object-in-space-before-subject. Without the parenthetic phrase, Being is reduced to what merely is ontical. Without the parentheses enclosing that phrase, Being is prone to be misread as but the *external cause* of the ontical. The parentheses support a reading of Heidegger's definition wherein we see that Being is at once the cause and the effect, the process and the product, the becoming present and the being present. Thus clarified, Heidegger's interpretation of the Anaximander fragment shows how the "things arise," how the beings become present, not just by random occurrence, not by their own hands alone, and not by the hands of what simply is other or absent. Instead the beings presence by virtue of an absence that is nonetheless more present to them than any ontical presence could possibly be!

Heidegger develops his own portrayal of the paradoxical nature of Being in the following way:

> [I]n Greek experience what comes to presence remains ambiguous, and indeed necessarily so. On the one hand, τα ἐόντα ["beings"] means what is presently present; on the other, it also means all that becomes present, whether at the present time or not. However, we must never represent what is present in the broader sense as the "universal concept" of presence as opposed to a particular case—the presently present—though this is what the usual conceptual mode of thought suggests. For in fact it is precisely the presently present and the unconcealment that rules in it that pervade the essence of what is absent, as that which is not presently present. (1946/1984, p. 35)

In other words, rather than presence and absence being treated dichotomously by regarding the latter as part of a more general concept of presence than the former, all "things present and absent are [to be seen as] gathered and preserved in *one* presencing" (p. 36). And: "Being, as the presencing of what is present…ambiguously names what is presently present and also what is not presently present; the latter, understood with regard to the former, means what is absent" (p. 37). Finally, what is presently present "comes to presence out of absence. This must be said precisely of whatever is truly present, although our usual way of representing things would like to exclude from what is present all absence" (p. 37). —Or, in denying Being, we might like to exclude from what is *absent* all *presence*, as Derrida sought to do in insisting that we should not call the "presence-absence of the trace…its ambiguity but rather its play (for the word 'ambiguity' requires the logic of presence, even when it begins to disobey that logic)" (1976, p. 71). Indeed, before rejecting the ambiguity of presence-absence in favor of his "*pure* absence" (*différance*), Derrida explicitly acknowledged "the ambiguity of the Heideggerian situation with respect to the metaphysics of presence" (1976, p. 22).

Now, the reflectivity of what is "presently present" entails boundaries and divisions. We know that, in the formulation of object-in-space-before-subject, both the objects and the space that contains them are infinitely divisible; moreover, all three terms of the classical formula are divided categorially from one another. In Heidegger's apprehension of Being, this reflectivity arises from a prereflective formative matrix in which categorial boundaries are not well established. What is "presently present," then, is the boundedness that issues from that which is *without* boundaries—from the *apeiron*. Here the intimate ambiguity of presence and absence translates as an ambiguity of boundedness and boundlessness, and we are led to the conclusion that *apeiron*, the boundless giver of boundaries, is none other than Being. Being, defined in terms of *apeiron*, is therefore *(the bounding of) what is bounded*, with the "bounding" implying the operation of a boundlessness, just as "presencing" implies the working of an "absence."

Heidegger approaches his correlation of Being with *apeiron* by way of considering Anaximander's term χρεών, conventionally translated as "necessary": "things arise...according to what is *necessary*...." Heidegger comments that while we "are accustomed to translate the word χρεών by 'necessity,'" which means "that which inescapably must be...we err if we adhere to this derived meaning exclusively" (1946/1984, p. 51). To render his own translation, Heidegger works with the etymology of χρεών:

> Χρεών is derived from χράω, χράομαι. It suggests ἡ χείρ, the hand; χράω means: I get involved with something, I reach for it, extend my hand to it. At the same time χράω means to place in someone's hands or hand over, thus to deliver, to let something belong to someone. But such delivery is of a kind which keeps this transfer in hand, and with it what is transferred. (pp. 51–52)

Heidegger next proceeds to relate χρεών to Being:

> If we firmly keep in mind that we must think the word [χρεών] within the Anaximander fragment, then it can only mean what is essential in the presencing *of* what is present [that is, in Being]....Τὸ χρεών is thus the handing over of presence which presencing delivers to what is present, and which thus keeps in hand, i.e. preserves in presencing, what is present as such. (p. 52)

The specific word that Heidegger ventures as the translation of τὸ χρεών is "usage" (p. 52). Acknowledging the "strangeness of the translation" (p. 52) when "usage" is understood in its "current, derived senses" (p. 53), he works with the root meaning of the term and arrives at the conclusion that, "'To use'...suggests: to let something present come to presence as such...to hand something over to its own essence and to keep it in hand, preserving it as something present" (p. 53). Thus, in the proposed "translation of τὸ χρεών, usage is thought as essential presencing in Being itself....[U]sage now designates the manner in which Being itself presences as the relation to what is present, approaching and becoming involved with what is present as present: τὸ χρεών" (p. 53).

Now, when "[u]sage delivers what is present to its presencing," it "bounds and confines what is present" so that that "which lingers awhile in presence...comes to presence within bounds" (p. 53). Usage then,

> [in] so limiting what is present, distributes boundaries. As τὸ χρεών it is therefore at the same time απειρον [*apeiron*], that which is without boundary, since its essence consists in sending boundaries…to whatever lingers awhile in presence….Anaximander is supposed to have said that whatever is present has its essential origin in what presences without bounds…απειρον. What presences without bounds…is not some present being but rather τὸ χρεών. (p. 54)

In his reflection on the portion of the Anaximander fragment that we have been considering—"But that from which things arise also gives rise to their passing away, according to what is necessary…."—Heidegger does not neglect the "passing away." The arising of things, their coming to presence, is hardly anything permanent. Instead, "What is presently present lingers awhile. It endures in approach and withdrawal. Lingering in the transition from coming to going: what is present is what in each case lingers. Lingering in transition, it lingers still in approach and lingers already in departure" (p. 37). Heidegger's words echo a line that continues to haunt me from a film whose name I have long forgotten: "Linger awhile, how fair thou art, how fair thou art." The poignant transience of this lingering bespeaks the rhythm of *apeiron* so well described by Nietzsche as his "eternally self-creating…eternally self-destroying…mystery world" (1895/1992, p. 106)—a serpentine world that Nietzsche ultimately could not stomach (Chapter Three). Only today (11 March 2001) did I learn the source of the line from the movie: It originated from the pen of J. W. Goethe, Nietzsche's alchemical predecessor. In sealing his pact with Mephistopheles, Goethe's Faust acknowledges that the *last thing* he can afford to say is "linger awhile, how fair thou art." For it is only by forging ahead, by moving onward and upward in his quest for God-like omniscience and omnipotence that Faust can avoid succumbing to the devil. To "linger awhile" thus would spell disaster for Faust; it would topple him from the heights of perfection toward which he has striven and cast him into that "black pit" in the bowels of the earth. Of course, Faust does finally succumb to Satan, just as Nietzsche's Zarathustra is pulled down by his own "spirit of gravity," his "devil and archenemy" (1878/1968, p. 268). In Heidegger's view, this is the consequence of failing to pay heed to the "constant danger that lingering will petrify into mere persistence" (1946/1984, p. 54). Unchecked by vigilance, petrifaction sets in and this one-sided denial of *apeiron* assures that sooner or later, one will confront *apeiron's* "Satanic" side and be petrified by her Medusan countenance. (Actually, while Faust's egological quest is ultimately vanquished and Mephisto tricks him into his grave, this is not the end of the story. In the closing scene, it is Mephisto who is tricked by a band of angels, and the deceased Faust's soul is then led to salvation, redeemed by the "Mother Goddess," as Adam McLean [1988, p. 40] puts it in his interpretation of Goethe's play. We may take this to signify *apeiron's* ultimate forgiveness of man's hubris.)

Heidegger goes on to extend his account of χρεών. He notes that, "What is thought as χρεών in the [Anaximander] fragment is the first and most thoughtful interpretation of what the Greeks experience in the name Moïρα....Gods and men are subordinated to Moïρα" (pp. 54–55). This allusion to the preeminence of Moïρα accords with the comment of philosopher W. T. Jones cited in Chapter One: "above the gods is fate, a blind, inscrutable 'will' to which even Zeus must yield" (1952, p. 29). Moïρα, the fate that rules the gods, was depicted as a mysterious goddess, or three such womanly figures, *the fates*. Recall Barrett's reference to them as "the Furies, the old goddesses of night and earth...[and] old Mother Earth herself" (1958, p. 277). Their "blindness" and "inscrutability" speak to their nonrational character. The philosopher F. M. Cornford agrees with Heidegger that this early notion of "fate" or "necessity" [χρεών] is decidedly different from what it means to us today. According to Conford, the "necessity" of Moïρα is not that which is "fixed, permanent, unalterable, knowable beforehand," but is "the indeterminate, the inconsistent, the anomalous, that which can be neither understood or predicted" (1957, p. 171). Of course, Heidegger would view Moïρα in more positive terms: she is *apeiron*, the boundless giver of boundaries who governs the "presencing of what is present" and thus constitutes the essence of Being. Putting it metaphorically, Moïρα is not merely "Bad Mother" (*in*determinate, *in*consistent, etc.), but "*Great* Mother." She is the Black Madonna.

To appreciate the uniqueness of Heidegger's approach, let us keep in mind the "gear" in which he operates in proceeding toward "the oldest fragment of Western thinking" (1946/1984, p. 13). Heidegger does not attempt to "turn around and face" the Anaximander fragment while moving in the "forward gear." That is, he does not function as an antiquarian, seeking to face the past as his object, and, in the process, turning his back on present and future concerns. Instead, he switches gears and moves *backward*. Thus moving "in reverse" toward the fragment, he obtains a sense of how our *current* situation—dominated as it is by objectifying consciousness—first arises from the pre-objective action of Being intimated early on by Anaximander.

In Heidegger's account of "early" and "late," we see how his reversal of gears confounds the splitting of past and future characteristic of linear chronology:

> The antiquity pervading the Anaximander fragment belongs to the dawn of early times in the land of evening [*Abend-Land*, the West]. But

> what if that which is early outdistanced everything late; if the very earliest far surpassed the very latest? What once occurred in the dawn of our destiny would then come, as what once occurred, at the last...that is, at the departure [or end] of the long-hidden destiny of Being. The Being of beings is gathered...in the ultimacy of its destiny....If we think within the eschatology of Being, then we must someday anticipate the former dawn in the dawn to come; today we must learn to ponder this former dawn through what is imminent. If only once we could hear the fragment it would no longer sound like an assertion historically long past. (1946/1984, p. 18)

When Heidegger suggests that we must "learn to ponder this former dawn through what is imminent," in effect he is proposing that our thinking *shift gears*. It is through such a reversal of orientation that, for the first time, *apeiron* can begin to be acknowledged and consciously accepted.

2. From Space-Time to "Time-Space"

The switching of gears is a shift into the *coagula*. In assuming this "reverse posture," it becomes clear that Being is indeed no mere point of departure for the stratospheric flights of the *cogito*, but is the whole embodied dimension from which the *cogito* first originates. We shall now attempt to sharpen our focus on *apeiron's* dimensionality. In so doing, let us return to a matter of crucial significance: the question of *space and time*.

We found in the previous section that Being is what first opens up the space in which objects are cast, and that this opening up is a presencing, a making present. To be "present" in the ontical sense is to be *here*, at this point in space. But ontical "presence" also refers to *now*, to the present moment in time. In the posture that obscures Being by reducing it to but a point of departure, time, as well as space, assumes a point-like character; when this occurs, time is apprehended as a mere sequence of extensionless "nows" (Aristotle's view of time). What happens to space and time with the ontological reversal that brings Being to light? These dimensions take on a certain "fullness" or "thickness." Heidegger explores the transformation in depth, particularly its temporal aspect. In his 1962 lecture, "Time and Being," he speaks of ontologically thickened temporality as "time-space" or "true time." According to Heidegger, far from being point-like, "true time proves to be three-dimensional" (1962/1972, p. 15).

Specifically, Heidegger tells us that, behind the appearance of an extensionless, point-like now is: (1) the *past*—which, though denied to us *as* present, nevertheless makes its presence felt in its own way; (2) the *future*—which is withheld from us in *its* own peculiar manner of making itself present; and (3) the present itself (the "presently present," as Heidegger referred to it in his earlier Anaximander meditation). Heidegger suggests that these three dimensions of time gather together and are held apart in every moment of lived experience. That is, they extend toward each other, yet, since they never merely collapse into one another to form a monotonic unity, the "space" of their intimate interplay remains open. The three dimensions of time determine each other; there is a "mutual giving to one another of future, past and present" (1962/1972, p. 14):

> [the future,] being not yet present, at the same time gives and brings about what is no longer present, the past, and conversely what has been offers future to itself. The reciprocal relation of both...gives and brings about the present....[Thus] future, past and present...belong together in the way they offer themselves to one another." (pp. 13–14)

Heidegger next observes that,

> With this presencing, there opens up what we call time-space. But with the word "time" we no longer mean the succession of a sequence of nows. Accordingly, time-space no longer means merely the distance between two now-points of calculated time....Time-space now is the name for the openness which opens up in the mutual self-extending of futural approach, past and present. This openness exclusively and primarily provides the space in which space as we usually know it can unfold. The self-extending, the opening up, of future, past and present is itself *pre*spatial; only thus can it make room, that is, provide space. (p. 14; emphasis added)

What Heidegger is saying is that—while the "dimensionality of time, thought as the succession of the sequence of nows, is borrowed from the representation of three-dimensional space" (p. 14)—when temporal dimensionality is thought in its *truth*, as the threefold mutual giving, then it is three-dimensional space that derives from *it.* Expanding on this, Heidegger asserts:

> [W]hat is germane to the time-space of true time consists in the mutual reaching out and opening up of future, past and present. Accordingly, what we call dimension and dimensionality...belongs to true time and to it alone. Dimensionality consists in a reaching out that opens up, in which futural approaching brings about what has been, what has been brings about futural approaching, and the reciprocal relation of both brings about the opening up of openness. Thought in terms of this

> threefold giving, true time proves to be three-dimensional. Dimension, we repeat, is here thought not only as the area of possible measurement, but rather as reaching throughout, as giving and opening up. Only the latter enables us to represent and delimit an area of measurement. (1962/1972, p. 14–15)

Of course, the "reaching throughout...giving and opening up" inherent in true time is the *action of Being*. To repeat, this action is *prereflective*; it takes place prior to the ontical activity of the subject with respect to its object once that subject has become detached. Therefore, to give the ontological difference its due, we must steadfastly avoid confusing the action of Being with that of ontical reflection. The former is what first makes the latter possible.

Consonant with Heidegger's analysis of dimension is the notion of *depth* advanced by the phenomenological philosopher Maurice Merleau-Ponty (1964). This idea provides us with a dialectical insight into dimensionality that, like Heideggerian thinking, permits us to understand the limits of classical space and to surpass them.

By way of introducing Merleau-Ponty's *depth* dimension, let us consider once more the traditional dichotomy between objects contained in space and their spatial container, or, as Plato put it, between "that which becomes [and] that in which it becomes" (1965, p. 69). A visible form "becomes," whereas that "*in* which it becomes" is "invisible and formless" (1965, p. 70). Whatever changes may transpire in the objects that "become," however they may be transformed, the containing space itself does not change. Indeed, for there to be change, there must be difference, contrast, dialectical opposition of some kind. But the point-elements that make up the classical continuum, rather than entailing opposition, involve mere *juxta*position. Unextended and thus devoid of inner structure, the elements of space possess no gradations of depth, no shading, texture, or nuance, no contrasts or distinctions of any sort. Instead of expressing the dialectical interplay of shadow and light, space itself is *all light*, as it were. A condition of "total exposure" prevails for the point-elements of the continuum, since these elements, having no interior recesses, must be said to exist solely "on the outside." All that can be said of the relations among such eviscerated beings is what Heidegger said: the points of classical space are "'outside-of-one-another'" (1927/1962, p. 481). So, rather than actively engaging each other as do the beings that are contained in space, the densely packed elements of the classical container sit inertly side by side, like identical beads on a string.

As a matter of fact, even though the beings that dwell in such a space can be described as "actively engaged," the quality of their interaction is affected by the context in which they are embedded: since the continuum is constituted by sheer externality, the relations among its inhabitants must also be external. Classical dynamics are essentially mechanistic; instead of involving a full-fledged dialectic of opposition and identity wherein beings influence each other from core to core, influence is exerted in a more superficial fashion. According to Bohm (1980), the mechanistic order of influence is one in which entities "interact through forces that do not bring about any changes in their essential natures...[they interact] only through some kind of external contact" (p. 173). We may say then that classical space contains dialectical process in such a way that it externalizes it, divesting it of its depth and vitality.

Now, while modernists like Bohm, Einstein, and Husserl decried the mechanistic attitude of classical science, while they did attempt to "dynamize" space and to "internalize" the interactions transpiring therein, the objectivist stance of modernism assured the ultimate failure of these efforts. In all cases, the classical formula reasserted itself at a higher level of abstraction, as we have seen, and this resulted in the eclipse of concrete dialectical process. Not so in the phenomenological approach of Merleau-Ponty. What Merleau-Ponty demonstrates is that the objective space appearing to contain dialectical process actually *originates* from it.

In his essay "Eye and Mind," Merleau-Ponty emphasizes the "absolute positivity" of traditional Cartesian space (1964, p. 173). For Descartes, space simply is *there*; possessing no folds or nuances, it is the utterly explicit openness, the sheer positive extension that constitutes the field of strictly external relations wherein unambiguous measurements can be made. Merleau-Ponty speaks of,

> this space without hiding places which in each of its points is only what it is....Space is in-itself; rather, it is the in-itself *par excellence.* Its definition is *to be* in itself. Every point of space is and is thought to be right where it is—one here, another there; space is the evidence of the "where." Orientation, polarity, envelopment are, in space, derived phenomena inextricably bound to my presence [thus "merely subjective"]. *Space* remains absolutely in itself, everywhere equal to itself, homogeneous; its dimensions, for example, are interchangeable. (1964, p. 173)

Merleau-Ponty concludes that, for Descartes, space is a purely "positive being, outside all points of view, beyond all latency and all depth, having no true thickness" (1964, p. 174).

Challenging the Cartesian view, Merleau-Ponty insists that the dialectical features of perceptual experience ("[o]rientation, polarity, [and] envelopment") are not merely secondary to a space that itself is devoid of such features. He begins his own account of spatiality by exploring the paradoxical interplay of the visible and invisible, of identity and difference, that is characteristic of true depth:

> The enigma consists in the fact that I see things, each one in its place, precisely because they eclipse one another, and that they are rivals before my sight precisely because each one is in its own place. Their exteriority is known in their envelopment and their mutual dependence in their autonomy. Once depth is understood in this way, we can no longer call it a third dimension. In the first place, if it were a dimension, it would be the *first* one; there are forms and definite planes only if it is stipulated how far from me their different parts are. But a *first* dimension that contains all the others is no longer a dimension, at least in the ordinary sense of a *certain relationship* according to which we make measurements. Depth thus understood is, rather, the experience of the reversibility of dimensions, of a global "locality"—everything in the same place at the same time, a locality from which height, width, and depth [the classical dimensions] are abstracted. (1964, p. 180)

Speaking in the same vein, Merleau-Ponty characterizes depth as "a single dimensionality, a polymorphous Being," from which the ontical Cartesian dimensions of linear extension derive, and "which justifies all [Cartesian dimensions] without being fully expressed by any" (1964, p. 174). The dimension of depth is "both natal space and matrix of every other existing space" (1964, p. 176).

Merleau-Ponty goes on to observe that primal dimensionality must be understood as *self-containing*. This is illustrated through a discussion of contemporary art, and, in particular, the work of Paul Cézanne: "Cézanne knows already what cubism will repeat: that the external form, the envelope, is secondary and derived, that it is not that which causes a thing to take form, that this shell of space must be shattered, this fruit bowl broken" (1964, p. 180). In breaking the "shell," one disrupts the classical representation of objects-in-space. Merleau-Ponty asks:

> [W]hat is there to paint, then? Cubes, spheres, and cones...? Pure forms which have the solidity of what could be defined by an internal law of construction...? Cézanne made an experiment of this kind in his middle period. He opted for the solid, for space—and came to find that inside this space, a box or container too large for them, the things began to move, color against color; they began to modulate in instability. Thus we must seek space and its content *as* together. (1964, p. 180)

The work of Cézanne is Merleau-Ponty's primary example of the exploration of depth as originary dimension. The foregoing passage describes Cézanne's discovery that primal dimensionality is not space taken in *abstraction* from its content—as in Cubist modernism, where the content is subordinated via objectification—but is the *unbroken flow* from container to content. It is in this sense of the internal mediation of container and content that Cézanne's depth dimension is *self-containing*. Like the serpent of old, it swallows itself.

Merleau-Ponty also makes it clear that the primal dimension engages embodied subjectivity: the dimension of depth "goes toward things from, as starting point, this body to which I myself am fastened" (1964, p. 173). In commenting that, "there are forms and definite planes only if it is stipulated how far from *me* their different parts are" (p. 180; italics mine), Merleau-Ponty is conveying the same idea. A little later, he goes further:

> The painter's vision is not a view upon the *outside*, a merely "physical-optical" relation with the world. The world no longer stands before him through representation; rather, it is the painter to whom the things of the world give birth by a sort of concentration or coming-to-itself of the visible. Ultimately the painting relates to nothing at all among experienced things unless it is first of all "autofigurative."....The spectacle is first of all a spectacle of itself before it is a spectacle of something outside of it. (1964, p. 181)

In this passage, the painting of which Merleau-Ponty speaks, in drawing upon the originary dimension of depth, draws in upon itself. Painting of this kind is not merely a signification of objects but a concrete *self*-signification that surpasses the division of object and subject.

In sum, the phenomenological dimension of depth as described by Merleau-Ponty is (1) the "first" dimension, inasmuch as it is the source of the Cartesian dimensions, which are idealizations of it; it is (2) a uroborically self-containing dimension, not merely a container for contents that are taken as separate from it; and it is (3) a dimension that blends subject and object concretely, rather than serving as a static staging platform for the objectifications of a detached subject. In other words, depth constitutes the dimensional structure of Being, the prereflective action of *apeiron* obscured by the ontical reflectivity of classical thinking. In realizing depth, we surpass the concept of space as but an inert container and come to understand it as an aspect of an indivisible cycle of action in which the "contained" and "uncontained"—object and subject—are integrally incorporated.

Now, since Merleau-Ponty's concept of depth as the "first dimension" does seem to parallel Heidegger's notion of "time-space" as primal dimensionality, we may ask whether a direct influence is in evidence here. The answer is yes. Although Merleau-Ponty makes no reference to Heideggerian temporality and its "three ecstases" (past, present, and future) in his essay "Eye and Mind," earlier, in *Phenomenology of Perception*, he quotes Heidegger extensively on this (1962, Chapter III-2). Here Merleau-Ponty arrives at the conclusion that human beings are essentially temporal creatures in Heidegger's sense of the term, and, moreover, that the mutual embeddedness and interpenetration of past, present, and future are inextricably related to the phenomenon of *perspective*: "Change presupposes a certain position which I take up and from which I see things in procession before me: there are no events without someone to whom they happen and whose finite perspective is the basis of their individuality. Time presupposes a view of time" (1962, p. 411). Of course, the dimension of depth of which Merleau-Ponty will later speak (in "Eye and Mind") is also the dimension of perspective: that wherein "I see things, each one in its place, precisely because they eclipse one another...." (1964, p. 180). (Merleau-Pontean perspective must be distinguished from its Cartesian counterpart, which is an idealization of perspectival experience that glosses over its inherent finitude.) What is the difference between Merleau-Ponty's later treatment of perspective and the earlier one that focuses on temporality as such? To answer this question, we return to our consideration of *Einsteinian space-time*. Let me take the liberty of repeating what I said in Chapter One:

In Einstein's special theory of relativity, space and time are united by replacing the hitherto separate measures of distance and duration with a new fundamental unit, the "space-time interval." A constraint placed on this unification is the need to distinguish unambiguously space-time intervals that are "space-like" from those that are "time-like." To put it roughly, although every space-time interval indeed combines space and time, any given interval must be "weighted" more by one than by the other. Essentially, two events in Einsteinian space-time are separated in a space-like way if the distance between them is great enough relative to their temporal disjunction that an observer could be present for the onset of both events only by traveling from one to the other at the velocity of light or greater. Einstein's theory assumes that no material body can travel that quickly. Events are separated by a time-like interval if the time elapsing between them is long enough relative to their spatial separation that an observer may be present for the onset of both events, but could not experience them as occurring simultaneously, since to do so again would require traveling at the prohibited velocity. We may apply this distinction to Merleau-Ponty's two approaches to perspective.

Let us say that, in "Eye and Mind," Merleau-Ponty's primary concern is with the "space-like" aspect of perspectival experience. It is here that he speaks of things "eclipsing" one another so that we cannot simply see both of them at once. From the Einsteinian standpoint, we would view the things in question as perceptual *events*, occurrences whose spatial separation is so great relative to their temporal disjunction (since the events would be taking place approximately at the *same* time) that I would have to travel at the velocity of light or greater to be present for the onset of both. In *Phenomenology of Perception*, Merleau-Ponty is chiefly interested in the "time-like" aspect of perspectival experience and therefore is focused on the question of the relationship of past, present, and future. This is the case of space-time events whose temporal component "outweighs" the spatial.

However, while Einstein's distinction may help us to conceptualize the difference between Merleau-Ponty's two analyses of perspective, it is important to realize that Merleau-Ponty's integration of space and time went considerably further than Einstein's. The space-time of Einstein is ontical whereas the "time-space" of Heidegger and Merleau-Ponty is ontological (or onto-phenomenological). To bring out the difference, we return to Figure 2 (p. 24), the Einstein-Minkowski space-time diagram. I noted in Chapter One that, like the 0,0 origin of Cartesian space, the here-now originating point of Einstein-Minkowski space-time plays a pivotal role in the classical intuition of object-in-space-before-subject. For, as the locus at which observation begins, the origin functions as the surrogate for the subject's "eyes," its point of perspective upon the world, its "window" upon the objects-in-space. We may contrast Einsteinian and onto-phenomenological approaches with respect to this origin. The former is geared to move *forward* from the here-now, to move out of and away from the origin in such a way that 0,0 is taken as but the dimensionless point of departure for the objective measurements and analyses carried out by the detached subject. In this ontical posture, the old subject-object dichotomy is unquestioningly maintained. What Merleau-Ponty achieves in his analysis of the "time-like" and "space-like" aspects of perspectival experience is, in effect, a *phenomenological reversal* of the Einsteinian orientation. "Shifting gears," Merleau-Ponty moves *backward* into the 0,0 origin, whereupon he finds that it is no mere transfer point in an external exchange between a detached subject and its object (as in the Einsteinian "space-time event"), but is the whole dimension of their internal relatedness. In so moving backward, the originating point "expands" or "thickens" into the sphere of action that Merleau-Ponty recognizes as Being. In sum, whereas Einstein moves away from Being thereby reducing it to a dimensionless point in the space-time continuum, Merleau-Ponty's onto-phenomenological reversal brings an awareness of the fullness of Being from which space-time derives. But let me be more specific about just what the reversal discloses.

For Merleau-Ponty, the present, the here-now, possesses a "dialectical thickness" not found in Einstein's account—or in any other account of presence in the long tradition of ontical posturing. Again, in "space-like" perspectival experience, "The enigma consists in the fact that I see things, each one in its place, precisely because they eclipse one another, and that they are rivals before my sight precisely because each one is in its own place" (1964, p. 180). This paradoxical interplay of the visible and invisible suggests that what is eclipsed is not simply elsewhere (see Fig. 2); it also is here. Although the presencing of elsewhere here defies the Einsteinian mandate that they be taken as externally related points, here and elsewhere do not collapse into a single point of immediate presence. Instead of constituting dimensionless points, here and elsewhere enact an internally mediated exchange that confers thickness on both. *Both*? Do we have two distinct terms, as in Einstein's analysis? Yes. But unlike what we find in Einstein, the two also are one.

We are led to a similar conclusion with respect to the "time-like" aspect of perspectival experience examined in Merleau-Ponty's earlier (1962) account. That work was based on Heidegger's interpretation of temporality, and we have seen that, for Heidegger, true time is no succession of dimensionless now-points but possesses "three-dimensional thickness." To be sure, the future is withheld from the present, and the past is denied to the present. Yet both nevertheless make their presence felt in the thickness of the now. Thus, owing to the mutual interpenetration of past, present, and future, dialectical thickness is not only found in the here, but in the now as well. Consequently, the future is not simply *after* the present, nor is the past simply *before* it—the proscriptions of the Einstein-Minkowski light cones to the contrary notwithstanding. What we have instead is the internal dialectic in which future and past—each in its own way—presences now, without merely collapsing into that now.

Note that it is when here and now serve merely as points of departure for measuring space-like and time-like intervals that spatial and temporal relations themselves are dualistically divided. In this ontical account, the light cones stretch out from the point of origin in such a way that the absolute future and absolute past are not only unambiguously set apart from one another, but from the elsewhere region of space-time as well. In the contrasting ontological approach—because elsewhere presences *here* and otherwhen presences *now*—elsewhere and otherwhen also are present to *one another*, each in its own way. We might imagine the light cones being retracted, reeled back into the here-now, yet not just deflating to a singular point at the center. In the paradoxical "thickening" of the here-now that would result, space and time themselves would overlap paradoxically. So the dialectic of here and elsewhere, of now and otherwhen, entails a dialectic of space and time themselves (see Abram, 1996, Chapter Six). In the interchange, Merleau-Ponty's depth dimension and Heidegger's "time-space" are inseparably linked. The phenomenological philosopher Sue Cataldi (1993) confirms this in her study of Merleau-Pontean depth when she asserts, "Merleau-Ponty continues to associate depth...with the thickness of a present holding a past and a future together" (p. 67).

The dialectical exchange in question is of course no "objective event," given that the subject also abides in the dynamic thickness of here-elsewhere/now-otherwhen. To repeat Merleau-Ponty's way of putting it, the primal dimension of depth "goes toward things from, as starting point, this body to which I myself am fastened" (1964, p. 173). In thus grounding the "I" in the body, the "I" ceases to be a high-flying Cartesian *cogito* poised only for the ontical forward thrust. A *backward* movement is involved in recognizing the "starting point" as densely embodied. A reversal takes place in which there is no longer a simple division between the subject's here-now and the things that lie elsewhere and otherwhen. Instead, an *internal circulation* of subject and object is set in motion, for the now-embodied subject—far from being the disengaged *deus ex machina* of Descartes and Einstein—finds itself down among the objects, is "one of the visibles" (Merleau-Ponty, 1968, p. 135), is itself always an object to some *other* subject, so that the simple distinction between subject and object is confounded and "we no longer know which sees and which is seen" (p. 139). In this regard, it is worth repeating what Merleau-Ponty said about the experience of artists like Cézanne. In precise contrast to Einsteinian experience,

> The painter's vision is not a view upon the *outside*, a merely "physical-optical" relation with the world. The world no longer stands before him through representation; rather, it is the painter to whom the things of the world give birth by a sort of concentration or coming-to-itself of the visible. Ultimately the painting relates to nothing at all among experienced things unless it is first of all "autofigurative"....The spectacle is first of all a spectacle of itself before it is a spectacle of something outside of it. (1964, p. 181)

In short, the phenomenological reversal by which we apprehend our "starting point" to be a whole dimension of lived embodiment discloses the circulation of Being, the serpentine flow of *apeiron*.

The failure of modern(ist) physics to deny resurgent *apeiron* was explored in Chapter One. In the ultimate confrontation with the radical discontinuity of the black hole, hope is dashed of imposing unity upon nature by containing it within the space-time continuum. What we can now see is that, in switching from space-time to the time-space of Heidegger and Merleau-Ponty, a kind of unity can indeed be achieved. The catch, of course, is that it is hardly the idealized unity to which the *cogito* has long aspired. Instead Heidegger and Merleau-Ponty offer us a paradoxical unity-in-diversity, one that is based on *accepting apeiron*, entering into her primal dimension.

Nevertheless, a further complication arises in our attempt to understand the meaning of primal dimensionality. In his late lecture on "Time and Being," Heidegger was not merely seeking a new approach to Being, to ontology; beyond that, he was attempting to surpass ontology itself. Thus, having associated Being with "true time"; and having spoken of the latter as the "three-dimensional" mutual giving of past, present, and future; Heidegger is prompted to ask, "from what *source* is the unity of the three dimensions of true time determined, the unity, that is, of its three interplaying ways of giving, each in virtue of its own presencing?" (1962/1972, p. 15; emphasis added). Heidegger answers the question in the following way: "[T]he unity of time's three dimensions consists in the interplay of each toward each. This interplay proves to be the true extending, playing in the very heart of time, the fourth dimension, so to speak—not only so to speak, but in the nature of the matter" (1962/1972, p. 15). In invoking a "*fourth* dimension," Heidegger is implying that the "dimension of interplay" at the heart of time must be taken as somehow constituting a dimensionality unto itself, rather than merely being a feature of three-dimensional time. Indeed, only thus can the "dimension of interplay" be seen as the *source* of three-dimensional time. Apparently then, there is a dimension that exceeds the three dimensions of time and constitutes their "source."

Heidegger avoids the metaphysical implication that the "fourth dimensional source" is a static realm or being by speaking of it in active terms, as a *giving*: "It gives time" (1962/1972, p. 16). But true time itself is a giving: it is that which "provides the space in which space as we usually know it can unfold" (p. 14). Accordingly, true time is what gives us the ability "to represent and delimit an area of measurement" (p. 15). Therefore, it seems we must regard the "fourth dimension" that gives true time as "the giving of a giving" (p. 16).

Later in his lecture, Heidegger attempts to go further with his portrayal of the "fourth dimensional It" that gives time and Being:

> In the sending of the destiny of Being, in the extending of time, there becomes manifest a dedication, a delivering over into what is their own, namely of Being as presence and of time as the realm of the open. What determines both, time and Being, in their own, that is, in their belonging together, we shall call: *Ereignis*, the event of Appropriation. (1962/1972, p. 19)

The ordinary meaning of the German word *Ereignis* is "event" or "occurrence." But Heidegger asks us to read this word in an unusually literalized fashion, to understand it as "making one's own." Thus we have the English translation of *Ereignis* as "Appropriation," from the Latin *ad*, to, and *proprius*, one's own. What this suggests is that the "fourth dimensional giving" entailed in Appropriation is not an imposition upon three-dimensional Being and time of something that is foreign to them, but is a giving to them of what is *their own*. To the extent that "fourth dimensional giving" is indeed fully committed to yielding over to Being what is truly Being's *own*, it seems the giving in question would not be so much an act of "fatherly condescension" that *determines* the "ownness" of Being, as a more nurturing gesture of support that encourages Being to give to *itself*. How does Being respond to this encouragement? Precisely by the presencing that gives an "area of measurement," gives space "as we usually know it," or, more generally, gives object-in-space-and-time-before-subject. Let us designate this giving of Being to itself as *self-Appropriation*.

Extending the metaphor, if "fourth dimensional" Appropriation is not a "fatherly" act, neither is the self-Appropriation of Being. By "fatherly" action, I mean that of the *deus ex machina*, the agent acting from above or outside. This action functions exclusively in its own interests, with the assumption that actor and acted upon are essentially separate. In other words, "fatherly" action describes the operations of the detached subject upon its objects. To be sure, this reflective subjectivity is *what* Being gives to itself. But the *giving* is not reflective. In so giving, Being does not take for itself what is other (as in "fatherly" action), but gives to itself what is its own. Being's mode of action is *pre*reflective, as we know; it gives itself reflective subjectivity in a manner that precedes the division of subject and object.

The point may be clarified, at least metaphorically, by considering the relationship of mother to child. Through her nurturing support, the mother facilitates the child's autonomy, encourages it to develop its unique way of acting, which is distinct from her own motherly care-giving. But, in thus giving to her child, the mother is not just acting selflessly on behalf of another. Because the mother's very being is intimately bound up with that of her child, by giving to it, she is also giving to herself. I suggest that this is the sense in which the prereflective giving of reflective subjectivity is Being's gift to itself—a process soon to be elucidated in considering the stages in Being's own development. I am proposing then that Being acts *maternally*, and the "child" to whom it gives birth is the subject. When we take into account Heidegger's sense of Appropriation, we have a tripartite relation. Operating in the manner of a "midwife," "fourth dimensional" Appropriation encourages Being's three-dimensional "motherly" self-Appropriation, and this, in turn, results in the birth of the "father," that is, of reflective consciousness.

The interpretation of Being as a "motherly" or "womanly" dimension is consonant, of course, with her portrayal as *apeiron*, the Black Goddess. And we have seen that Heidegger himself lends support to this view in an earlier essay. In Heidegger's translation of the Anaximander fragment, the "presencing of what is present" that constitutes the action of Being does not so much occur according to what is "*necessary*," as the word χρεών is conventionally translated, but according to "usage." Again, "'To use'... suggests: to let something present come to presence as such...to hand something over to its own essence and to keep it in hand, preserving it as something present.... [U]sage now designates the manner in which Being itself presences as the relation to what is present, approaching and becoming involved with what is present as present" (1946/1984, p. 53). Evidently, to "use" is to "*appropriate*," for we have just found that, in Heidegger's subsequent lecture, "*Ereignis*, the event of Appropriation," is defined as "a dedication, a delivering over into what is their own...of Being as presence and of time as the realm of the open" (1962/1972, p. 19). What the later lecture establishes is the *twofold* nature of usage or Appropriation: there is a "giving of a giving" (p. 16) in which the "fourth dimensional It" gives Being, and Being, in turn, gives "space as we usually know it" (p. 14). And the *earlier* lecture, for its part, more explicitly intimates the *womanly* character of Appropriation. Here the appropriative action of Being is taken as the action of *apeiron*, and *apeiron* is related (albeit somewhat indirectly) to Μοῖρα, the goddess to whom "Gods and men are subordinated" (1946/1984, p. 55).

3. Ontogeny and the Call to Being

The God of patriarchal tradition is deemed immutable. He is said to be an unmoved mover, a creator who is himself exempted from change by virtue of his detachment from his creatures. The feminine counterpart of the God cannot function in such a "pristine" fashion. She must "get down and dirty." That is, she must participate in the creative process with the whole of her body, and therefore cannot help but be affected by it. To speak as we have of Being's self-Appropriation, of her "midwife-assisted self-giving," is to intimate that the apeironic Goddess indeed undergoes a bodily transformation: *She gives birth to herself.* Presently, we shall begin to consider this developmental process in a systematic way. In exploring the history of Being, highlights of the historical account provided in previous chapters will be reprised, this time from an onto-phenomenological vantage point.

The term "ontogeny" is commonly defined as the development of an individual being, with "being" tacitly understood in the *ontical* sense. In the present context, I will capitalize the word "Ontogeny" to indicate that here we are speaking *ontologically*, dealing with the development or individuation of Being itself. Before proceeding with this, let us first take a closer look at the *sense* in which Being develops.

Heidegger often spoke of the "forgetfulness" of Being and of the need for its "remembrance." Echoed in this characterization is the Hegelian theme of Absolute Spirit's self-alienation and eventual self-return. Yet changes in the *awareness* of Spirit or Being do not necessarily imply a core transformation of Being as such. In the classical tradition, where epistemology and ontology—knowing and Being—are kept strictly separate, a change in the former holds no necessary implications for change in the latter. Accordingly, Being could "forget its true self" without changing that "truth" one iota. We know, in fact, that the transformation of Being per se is expressly precluded in classical thinking. In the categorial distinction set forth in Plato's *Timeaus*, it is clear that Spirit or Being (the fatherly creator or Demiurge) is not "that which becomes" (1965, p. 19); rather, it is *matter* that becomes, *bodies* that are subject to transformation. Permitting Being to "forget itself" does nothing to alter this. Indeed, what Being's "self-forgetfulness" means in this tradition is precisely that it has lost awareness of its own absolutely changeless character, an event presumably occasioned by the "fall into matter," the somnambulistic descent of Being into the relative world of transitory bodies.

In associating Being with pre-Platonic *apeiron* and *moira*, Heidegger does implicitly question the Platonic tradition of disembodied Spirit. This challenge becomes more explicit at the hands of Merleau-Ponty. For him, Being is not rarefied Spirit but is *flesh*, the "flesh of the world." Merleau-Ponty declares that this notion of flesh is unprecedented in Western thought. "[T]here is no name in traditional philosophy" for it (1968, p. 139) because, with it, the most basic categories of the classical tradition are transgressed:

> The flesh is not matter, in the sense of corpuscles of being...is not mind, is not substance. To designate it, we should need the old term "element," in the sense it was used to speak of water, air, earth, and fire, that is, in the sense of a *general thing*, midway between the spatio-temporal individual and the idea, a sort of incarnate principle that brings a style of being wherever there is a fragment of being. The flesh is in this sense an "element" of Being. (1968, p. 139)

If we take Merleau-Ponty at his word, if incarnate Being is a hybrid form that has as much in common with matter as it does with spirit, then Being surely cannot be exempted from change. Matter (earth, air, water, etc.) is in process; it undergoes transformation; and *living* matter develops, passes through stages of biogenesis. I propose then that we regard Being as possessing the character of a living organism—though not a finite particular organism (a "spatio-temporal individual"), to be sure. Instead, Being is a generic organicity, a whole dimension of life, a life*world.* Nonetheless, that world develops. We shall soon begin to detail the unique nature of Ontogeny that distinguishes it from development as ontically conceived. For the moment, suffice it to say that apeironic Being does develop.

Does Merleau-Ponty go so far as explicitly to equate the "flesh of the world" with *apeiron*? Not that I know of. Still, his portrayal of Being as "brute or wild" (1968, p. 170), as involving "polymorphism" and "promiscuity" (p. 253), seems to intimate the working of *apeiron*. The phenomenological philosopher Gary Madison (1981, p. 244) picks up on the connection in his commentary on the pre-Socratic roots of Merleau-Ponty, and the relationship is confirmed in philosopher Ernest Sherman's paper on "Merleau-Ponty and the Trickster" (1993). As noted in the last chapter, the mythological Trickster undoubtedly betokens *apeiron*, a fact that Sherman establishes up front in an epigraphic citation of Karl Kerényi: "'[Trickster's] nature, *inimical to all boundaries*, is open in every direction'" (1993, p. 103, italics added). Sherman then guides us through the serpentine ambiguities of the Trickster found in Greek mythology, by way of demonstrating that this same Mercurial force guides the thinking of Merleau-Ponty.

Now, it is clear from Merleau-Ponty's dialectical description of the flesh that, while the flesh is surely not *pure* spirit, the "spiritual" aspect of the ontological hybrid deserves equal emphasis. If Being is not just disembodied mind or subjectivity, neither is it mere objective fact. In other words, there is an inner dimension to Ontogeny; Being's growth and development is "*psycho*-physical," not merely physical.

With respect to gender, the hybrid character of Merleau-Pontean Being suggests the *hermaphroditic* face of the Goddess. Though I have thus far used feminine descriptors to characterize *apeiron*, hardly can "she" be regarded as feminine *as opposed to* masculine, since to do so would be to define "her" one-sidedly, in the categorial terms of patriarchy. In this connection it is interesting that the alchemical literature is replete with images of the hermaphrodite, with the Spirit Mercurius itself often being so depicted. Moreover, a recurrent theme of alchemy is that of the *hierosgamos*, the "mystical marriage" of the "king and queen." Therefore, to avoid gender splits and stereotypes, let us take our cue from alchemy and say that, if Ontogeny is psycho-physical, it is also "femino-masculine." Henceforth, when I portray individuation as a "motherly" or "maternal" process, these terms will be understood in the non-dualistic, dialectical sense of maternity, not in the stereotypically "feminine" sense. This does not mean that apeironic Being is gender *neutral*. Being is not androgynous but hermaphroditic, and we are going to discover that *apeiron* is a "hermaphrodite" in the specific sense that, rather than being "all woman," s/he contains within her the seed of a masculinity whose development is inextricably entwined with her own.

We are now prepared to launch a detailed exploration of Ontogeny. What we are about to see is that Being does "forget itself," then "remembers itself," in the course of its individuation. It is just that these changes are not epistemic alone but are changes in Being per se, since Being is *living flesh*. Let us proceed then to sketch out the stages of onto-dimensional development.

What is the first stage of Ontogeny? In Chapter One, I indicated that science's quest for unity or invariance can be traced all the way back to the pre-scientific development of object constancies in early childhood. Prior to the stabilization of these constancies, objects are seen to *change* in immediate relation to changes in the observer's perspective. Then, as experience becomes more abstract, the reality of this linkage between observer and observed, subject and object, tends to be denied; we come to say that the object only *appeared* to change, that, in actuality, it is only our perspective that changes. By thus detaching the observational framework from that which is observed, the notion of invariance or symmetry arises. In its essence, the scientific enterprise entails a furtherance of this detachment process.

Nevertheless, to understand the developmental process in Ontogenetic terms, it will not suffice to go back to the young child's concrete linkage with what she observes. We must go still further backward into infancy. At the very outset of development, there is no "infantile observer," no immature Cartesian subject, no individual who is epistemically linked to its objective environment but ontologically detached from it. Ontologically—Ontogenetically—what we are dealing with at the outset is the opening phase in the development of Being itself, a stage in which the subject has not yet been separated from its object, that wherein the projection of object-in-space-before-subject has not yet occurred. From the ontological standpoint, we can appreciate the basic flaw in ontical analyses of the infantile state of affairs: the presupposition of object-in-space-before-detached-subject is applied to a situation in which it is inappropriate. In the apeironic turbulence prevailing before the first object constancies, it is not the case that objects are seen by an "infantile subject" to change in relation to changes in its perspective. It is that the "infant" has no well-defined perspective (Sheets-Johnstone, 1990, p. 4), that there *is no infant*, in the sense of a freestanding subject (immature or not) before whom objects are projected. Ontologically speaking, what we have in infancy is not so much an epistemic fusion of subject and object, with the two terms taken as constituting already well-formed categories of being. Instead there is a lack of differentiation of these orders of being.

Let us say that, in infancy, embryonic subjectivity "rests in the womb" of Being. Then three-dimensional Being responds to the encouragement it receives from the Appropriative urgings of the "fourth dimensional midwife." The self-Appropriative labor of Being thus begins, and, when the period of gestation is completed, Being can give to itself what is its own by sending forth from its womb the ontical orientation of object-in-space-before-subject. Note that, because this "child" is Being's gift to *itself*, the child's development is not only, or even primarily, what is at stake here. Of foremost concern is the "mother's" own development. By giving itself subject-object differentiation, Being *clarifies* itself.

It may be useful to distinguish ontical self-clarification from the self-clarification of Being. In the former case, what is clarified is a specific matter, a finite particular being assumed to be already projected before a reflecting subject. In contrast, the prereflective self-clarification of Being that constitutes its self-Appropriation is the *primordial self-orienting action* that first makes possible clarification of the ontical kind. In the initial phase of Ontogeny, what is unclear is not some particular object, but the distinction between subjectivity and objectivity as such. We could say that, in this stage, Being is disoriented; it lacks direction. Then, in the ensuing stage, Being commences to clarify the relationship between subject and object, to differentiate them by conferring a direction upon their interaction. With the projection of object-in-space-before-subject, awareness is now oriented to function in a single direction: *from* the subjective ground of reflection *to* the objects that are cast before the reflecting subject. Of course, this irreversible from-to gearing describes the "forward" propensity of ontical thinking, the disposition that reduces Being to but a point of departure, as discussed above. Obscured by the subject-to-object movement of reflection is the prereflective self-Appropriation of Being that *first opens up* the field of objectification, and, in the same stroke, first gives a subject who detachedly reflects upon the objects in this field. Initially disoriented Being thus affords itself direction by transforming itself into intentional subjectivity, subjectivity-in-reflection-upon-its-field-of-objectification. In the process of clarifying itself in this manner, Being *conceals* itself. It facilitates its own development by moving away from itself. In this second stage of Ontogeny, Being gives itself reflective clarity by means of a prereflective process that is itself excluded from the giving.

It is important to recognize that Being's self-giving is not deliberate or even conscious in this stage of its development. If it seems strange that Being could clarify itself and conceal itself without being consciously aware of doing so, then we must be clearer about the difference between prereflective maternal action and the *father's* way of acting. It is the latter that entails conscious acts of deliberation carried out by a reflective subject. Intentional subjectivity is what Being *gives*, but the giver itself is no such subject. Perhaps we can characterize the prereflective action of apeironic Being as "instinctive": "impelled by an inner or animating agency; hence, imbued; filled; charged; as, a poem *instinct* with passion" (*Webster's*, 1975, p. 436). The *modus operandi* here is "natural, voluntary, spontaneous…impulsive" (*Webster's*, 1976, p. 951). What we must keep in mind is that a*peiron* is no coolly deliberative *cogito*.

In terms of cultural history, the second stage of Ontogeny was significantly advanced with the transmutation of consciousness that took place at the time of the Renaissance, a transformation subsequently cemented in the dualistic philosophy of Descartes (see Chapter One). It was not simply that mind and body were sharply divided at this time, but that they were divided in such a way that mind "ascended to the stratosphere" while body stayed "down on the ground." That is, what Descartes's work promoted was the *abstraction* of the subject, its detachment from the world of concrete bodies and happenings. By the same process, the bodies were relegated to the status of being *mere objects*; they were to be treated as merely factual, as having—in the words of Heidegger—"no concealed qualities, powers, and capacities...[but being] only what they *show* themselves as" (1962/1977, p. 268). So—while the subject or self was largely withdrawn from the realm of concrete things, people, and events, contracting to the interior point of perspective, the "I" imagined to lie inside one's forehead, behind one's eyes—nature, for its part, was stripped of its subjectness, becoming mere object. No longer lived in its fleshy, immediate presence, nature became that which was re-presented (symbolized) within the "uniform space-time context" (Heidegger, 1962/1977, p. 268), the spatial framework in which all bodies could be precisely located, probed and experimented upon, submitted to mathematical analysis, and, ultimately, shaped and manipulated through the enterprises of science and technology. The objects of nature thus were laid open, fully exposed, extended for scrutiny (they were *res extensa*), whereas the subject—being *un*extended in Descartes's definition of mind or spirit—was closed to view, essentially anonymous. It was in this way that the forward orientation of consciousness was sharpened and enhanced, made completely irreversible. Following the Renaissance, the subject—having become detached and disembodied, thus rendered anonymous—was not to *be* known; *it* was to do the knowing. Was this subject-object dichotomy an errant deviation from Being? On the contrary, it was Being's *gift* to itself, its manner of self-clarification.

Being's self-Appropriation was carried out in oblivion of Being up until the middle of the nineteenth century. Yes, at least as far back as the time of early Greece, but especially from the time of the Renaissance, Being gave itself what is its own by giving itself intentional subjectivity. But just this act of giving is what was obscured most assiduously, thereby maintaining the *cogito's* ontical posture. To repeat, Being, in its self-Appropriation, was poised to move *away* from itself. What happened in the middle of the nineteenth century? The process of self-Appropriation approached its culmination and limit, whereupon apeironic Being began to make its presence felt in an unprecedented way.

The resurgence of *apeiron* at mid-nineteenth century was a major theme in Part One of this book. We saw how *apeiron* was manifested in science (Chapter One) and in the broader culture (Chapter Two). I referred to the new era ushered in by these manifestations as the "Age of Light," since the phenomena in question were largely related to electromagnetism. The Michelson-Morley experiment was a prime example. This research was to corroborate the widely held view that light could be treated as a mechanical disturbance in the hypothesized luminiferous ether—in other words, that it could be treated as an *object-in-space*. To the amazement of the scientific community, the confirmation failed. We explored the underlying implications of this finding in Chapter One and reached the conclusion that light cannot be treated as an object open to the scrutiny of a subject that stands apart from it, because light is not simply something that we see, but is that *by which* we see; as such, light entails an inseparable blending of subject and object. In order to grasp this from the Ontogenetic standpoint, let us rewrite Heidegger's formula for Being—the "presencing of what is present"—so that it reads: the *lighting of what is lit*. We are then able to say that the subject-object split had been maintained during the second stage of Ontogeny by limiting attention to the objects that are lit for the reflecting subject; under this restriction, the *prereflective lighting process* had been consigned to darkness. In this way, Being had clarified itself through the reflective consciousness of its "child," while concealing its own light. Then, with the advent of the Age of Light at mid-nineteenth century, came the indication that Being's self-concealment was reaching its end, and that a new stage of Ontogeny might be at hand.

The connection between Being and light is approached from a different departure point in Heidegger's essay, "The End of Philosophy and the Task of Thinking" (1964/1977). Heidegger starts off by interpreting the mid-nineteenth century watershed as the "end of philosophy." Philosophy ends with the coming to dominance of scientific positivism—an historical development whose onset coincides with the dawning of the Age of Light. But didn't the Michelson-Morley experiment represent a *failure* of science and concomitant resurgence of *apeiron*? It was hardly that simple, as we know. While the research of Michelson and Morley did mark the failure of *classical* physics, at the same time it provided the occasion for creating a *new* physics in which *apeiron* was newly repressed; in the process, new heights of precision were achieved. Heidegger's point is that this modern science is what philosophy becomes. Philosophy thus "ends," not as a mere termination, but as a culmination of the scientific project. Heidegger states accordingly: "The end of philosophy proves to be the triumph of the manipulable arrangement of a scientific-technological world" (1964/1977, p. 377). However, in philosophy's realization of its "most extreme possibilities" (1964/1977, p. 375), Heidegger intimates that one possibility may have been overlooked:

> But is the end of philosophy in the sense of its evolving into the sciences also already the complete actualization of all the possibilities in which the thinking of philosophy was first posited? Or is there a *first* possibility for thinking apart from the *last* possibility which we characterized (the dissolution of philosophy in the technologized sciences), a possibility from which the thinking of philosophy would have to start, but which as philosophy it could nevertheless not experience and adopt? (1964/1977, p. 377)

If there were such a "first possibility for thinking"—one that was unrealizable throughout the history of philosophy but can be broached now that philosophy has ended—then the essential task of thinking would be to think that possibility.

Heidegger next notes that both Hegel and Husserl addressed themselves to the task of thinking in their phenomenological "call 'to the thing itself'" (p. 383). "From the perspective of Hegel and Husserl," says Heidegger, "the matter of philosophy [the "thing itself"] is subjectivity" (p. 383). To be sure, what concerned Hegel and Husserl in their "call," what they were trying to think, was the "presentation" of subjectivity, the way it "becomes present" (p. 383). In other words, rather than taking subjectivity for granted or treating it as an object in the old mechanistic sense, Hegel and Husserl sought to think subjectivity into its own, "to its ultimate originary givenness,...to its own presence" (p. 383). This means that they sought to think the *Being* of subjectivity. Nevertheless, in focusing on *subjectivity*, in making Being as subjectivity the matter of philosophy rather than Being *as Being*, something remains *un*thought:

> [W]hat remains unthought in the matter of philosophy as well as in its method? [Hegel's] [s]peculative dialectic is a mode in which the matter of philosophy comes to appear of itself and for itself, and thus becomes present. Such appearance necessarily occurs in some *light.* Only by virtue of light, i.e., through brightness, can what shines show itself, that is,

> radiate. But brightness in its turn rests upon something open, something free, which might illuminate it here and there, now and then....Wherever a present being encounters another present being...but also where, as with Hegel, one being mirrors itself in another speculatively—there openness already rules, the free region is in play. Only this openness grants to the movement of speculative thinking the passage through what it thinks. (1964/1977, pp. 383–384; italics mine)

Evidently then, in the phenomenological initiatives of Hegel and Husserl that attempt to think Being as subjectivity, what remains unthought is Being as such, Being as *radiance*, along with the opening that grants this radiance. This conclusion is confirmed by a subsequent remark of Heidegger's, made when he is asking why openness has been misunderstood: "Is it because man's ecstatic sojourn in the openness of presencing is turned only toward what is present and the [subjective] presenting of what is present [as with Hegel and Husserl]? But what else does this mean than that presence as such, and together with it the opening granting it, remain unheeded?" (p. 390).

The light of Being as such and the opening that makes it possible define "the first possibility for thinking." Heidegger informs us that it was not only the thinking of Hegel and Husserl that fell short of this "first thinking." Never in the history of philosophy has such a thinking occurred, although, at the dawn of philosophy, a glimmer of it was given in a fragment of Parmenides: "'but you should all learn...the untrembling heart of unconcealment'" (1964/1977, p. 387). Heidegger translates "unconcealment" (*alētheia*) as *opening*. So the task of thinking at the end of philosophy is to think Being and its unconcealment.

After discussing the etymological and factual relationships between the words "light" and "open" (in German the adjective *licht* can connote both), Heidegger states: "It is necessary for thinking to become explicitly aware of the matter here called opening....[T]he phenomenon itself, in the present case the opening, sets us the task of learning from it while questioning it, that is, of letting it say something to us" (p. 385). With this observation, Heidegger does nothing less than redefine phenomenology. No longer is phenomenology's call to the "thing itself" merely a call to the presencing of subjectivity; we are now called to something more primordial, to a "primal phenomenon" (p. 385): to openness or *alētheia*, and to the light of Being that it grants.

Perhaps the reader will recognize an underlying structural similarity between "The End of Philosophy" and two essays of Heidegger's that we studied earlier: "The Anaximander Fragment" and "Time and Being." Let us first compare "The End of Philosophy" with "Time and Being." The two works appear to entail the same three levels of analysis: the ontical, the ontological, and that which goes beyond ontology. In being "turned only toward what is present," we maintain the forward directed, subject-to-object, ontical orientation in which objects are set up in ordinary space (the "area of measurement") for the maneuverings of a detached subject. This ontical stance is first made possible by the self-Appropriation of Being, which, in "The End of Philosophy," is associated with *light*. Again, "Only by virtue of light...can what shines show itself, that is, radiate" (1964/1977, p. 383). Above I described Being's self-Appropriation of object-in-space-before-subject as a self-*clarification*; indeed, in clarifying something for oneself one sheds light on it (the words "clarity" and "light" obviously bear a close relationship; see Heidegger, 1964/1977, p. 384). Heidegger also associates light with the old Greek word, *energeia*: "that which in itself and of itself radiates and brings itself to light" (p. 385). Being's self-Appropriation then, is the energetic act of clarifying itself, bringing light to itself, giving to itself the radiance that first makes possible what we see, that first opens up space "as we usually know it" (Heidegger, 1962/1972, p. 14). In studying "Time and Being" we additionally found that Being's gift to itself, its "motherly" self-Appropriation, is encouraged by an other-dimensional "midwifely" act of Appropriation that goes beyond Being, that exceeds ontology. Evidently, the midwifely giving that Heidegger names *Ereignis* in "Time and Being" corresponds to what he calls *alētheia* in "The End of Philosophy": the unconcealment or opening "which first grants Being" (1964/1977, p. 387). Explicit confirmation of this comes at the very end of the former essay. Having meditated at length upon Appropriation, Heidegger acknowledges that its meaning remains elusive. And yet, it "is not even anything new, but the oldest of the old in Western thought: that ancient something which conceals itself in *a-letheia*" (1962/1972, p. 24).

Of course, "the oldest of the old in Western thought" is also reflected in "The Anaximander Fragment." Reexamining this earlier essay, we again find the ontical, ontological, and that which goes beyond ontology. The reference comes following Heidegger's contention that, to grasp Parmenides and Heraclitus, we must first understand Anaximander. In order to do the latter,

> we must first cross over to what that always unspoken word, ἐόν, ἐόντα, εἶναι [being, beings, to be] says: It says: presencing into unconcealment. Concealed in that word is this: *presencing brings unconcealment along with itself.*
>
> What is present is that which, whether presently or not, presences in unconcealment....Ἀλήθεια...belongs to the essence of Being.... (1946/1984, p. 55)

So, while Heidegger attributes the notion of Ἀλήθεια to Parmenides in "The End of Philosophy," we see from the earlier essay that it actually goes back to Anaximander. "What is present is that which…presences in unconcealment," says Heidegger. This may be the simplest way of expressing the threefold relation common to all three essays: that among beings, Being, and its opening. To underscore the intimacy of presencing and unconcealment, Heidegger says, "Unconcealment itself is presencing. Both are the Same, though they are not identical" (1946/1984, p. 55). Such is the paradoxical entwinement of three-dimensional motherly self-Appropriation and the other-dimensional midwife that Appropriates on her behalf.

Another parallel is worth noting. In his 1964 essay, Heidegger tells us that it is only at the "*end* of philosophy" that we are able to consider "the *first* possibility for thinking." We found the same reversal of beginning and end in Heidegger's 1946 lecture:

> The antiquity pervading the Anaximander fragment belongs to the dawn of early times in the land of evening [*Abend-Land*, the West]. But what if that which is early outdistanced everything late; if the very earliest far surpassed the very latest? What once occurred in the dawn of our destiny would then come, as what once occurred, at the last…that is, at the departure [or end] of the long-hidden destiny of Being. The Being of beings is gathered…in the ultimacy of its destiny….If we think within the eschatology of Being, then we must someday anticipate the former dawn in the dawn to come; today we must learn to ponder this former dawn through what is imminent. (1946/1984, p. 18)

What is "today imminent" is the *end of philosophy*. This "departure" thus calls us back to the beginning. That is, it calls us back to *Being*.

In view of the foregoing, we are able to appreciate better the Ontogenetic significance of the developments that occurred in the middle of the nineteenth century. The Age of Light is the *call back to Being*. In the mechanical age that was part of the second stage of Ontogeny, Being oriented itself, illuminated itself, gave itself reflective clarity in such a way that its own prereflective act of giving was shrouded in shadow. With the nineteenth century completion of this stage came the "lightning flash" that called us backward to something still ahead. Now, for the first time, three-dimensional Being was to *acknowledge* the giving of light to itself (along with the gift it had been given by the other-dimensional "midwife"). At bottom, this was the meaning of the Michelson-Morley experiment: We were no longer merely to think what is lit but light itself; that is, we were to think the light of Being. The "task of thinking" at this juncture then was to "switch gears"; to reverse the prevailing orientation of thinking that moves out of and away from Being; to move backward *into* Being, thereby drawing back in the radiance that prereflectively had been emitted, withdrawing its projection. In our present day—having spent more than a century avoiding the task of thinking via the excursion into modernism that has now reached its postmodern dead end—the task remains.

To put it differently, I suggest that the task of thinking in the third stage of Ontogeny is to think *proprioceptively* (see Rosen, 1997). Etymologically, to *per*ceive is to "take hold of" or "take through" (from the Latin, *per*, through, and *capere*, to take), and to *con*ceive is to "gather or take in." These activities correspond to the ordinary *from-to*, forward gearing of ontical consciousness. The term "proprioceive" is from the Latin, *proprius*, meaning "one's own." Literally, then, proprioception means "taking one's own," which can be read as a taking of self or "self-taking." The term finds its most common usage in physiology where it signifies an organism's sensitivity to activity in its own muscles, joints, and tendons. But Bohm (in Rosen, 1994) spoke of the need for "*proprioceptive thought*" (p. 229), which he viewed as a certain kind of meditative act wherein "consciousness...[becomes] aware of its own implicate activity, in which its content originates" (p. 232; see also Sundararajan, 2002). Years earlier, the social psychiatrist Trigant Burrow spoke similarly of the need for human beings to gain a proprioceptive awareness of the organismic basis of their divisive symbolic activity (Galt, 1995). What I am proposing is that events in the middle of the nineteenth century call *Being* to function proprioceptively, for they do not merely call for Being's awareness of what it lights (namely, object-in-space-before-subject), but of its own lighting process.

We already know that, in Heidegger's literal usage, the word "Appropriation" means "to one's own," from the Latin *ad*, to, and *proprius*, one's own. Expanding on this, I have suggested that stage two of Ontogeny involves both an extra-dimensional Appropriation on behalf of three-dimensional Being and an act of *self*-Appropriation in which Being gives to itself what is its own. What I presently suggest is that, in stage three, Being is called back to itself in such a way that it must *receive* these gifts. That is, it must engage in Proprio-*ception*. With this realization of the "first possibility for thinking," the circle would be closed, the process of giving completed in earnest.

If we regard Proprioception as an acknowledgment of the Appropriation that has taken place, a reciprocation of the gift that has been given by "saying thanks" for it, then we are led once again to Heidegger's reflections on the sort of thinking now called for in the call to Being. Noting the etymological consanguinity of "thinking" and "thanking," Heidegger (1954/1968) says:

> What gives us food for thought ever and again is the most thought-provoking. We take the gift it gives by giving thought to what is most

> thought-provoking. In doing so, we keep thinking what is most thought-provoking. We recall it in thought. Thus we recall in thought that to which we owe thanks for the endowment of our nature—thinking. As we give thought to what is most thought-provoking, we give thanks. (pp. 145–146)

What is it that is "most thought-provoking"? To what do "we owe thanks for the endowment of our nature"? It is clear that, for Heidegger, it is Being. Therefore, in thinking Being, we think Proprioceptively; we thankfully receive what has been Appropriated, what has been given over to what is our own.

Nevertheless, Heidegger acknowledged that, "Not one of us here would presume to claim that he is even remotely capable of such [Proprioceptive] thinking, or even a prelude to it. At the very most," admitted Heidegger, "we shall succeed in preparing for it" (1954/1968, p. 146). The profound difficulty of heeding the call to Being should come as no surprise, given what we learned in Part One. There we discovered that the mid-nineteenth century challenge to classical thinking was far from a mere intellectual conundrum. The questions raised about the primacy of the classical formula led to underlying doubts about the whole system of external relations upon which human identity itself had come implicitly to rely for countless centuries. To repeat the conclusion of Elizabeth Grosz, "A stabilized body image,...a consistent and abiding sense of self and bodily boundaries, requires and entails understanding one's position vis-à-vis others, one's place at the apex or organizing point in the perception of space" (1994, p. 48). More strongly put, the life of the self-standing subject, its very bodily existence, essentially depends on it standing in external relation to objects appearing before it in space. In the development of the individual, this relationship dates back to infancy; in phylogeny, it originates at the dawn of human culture. And this was the relationship that was called into question at mid-nineteenth century.

Perhaps it is not enough to say that the momentous events of the nineteenth century mark the "end of philosophy." More than that, they signify the end of *ontogeny*. I say this because ontogeny is the history of being-in-space, of the seemingly autonomous subject who has the "capacity to take itself as an object" (Grosz, 1994, p. 32). We may view ontogeny itself as the intermediary stage of *O*ntogeny, the stage before which and after which human experience is not governed by the ontical posture of object-in-space-before-subject. It is in the ontical stage that Being clarifies itself via the objectifications of its "child." The light Being thus gives (while concealing the giving process) illuminates the empty space in which vast networks of things, people, and events now can form projectively. The light that shines on these assemblages of beings is mirrored back to Being in various patterns of diffraction, creating the host of individual perspectives that constitute the vantage points of "free-standing" subjects or selves. When Being clarifies itself by concealing itself in this way, it looks in the mirror of space and sees only *beings*, thereby confirming itself in the guise of the ontical.

So it was that, in the second stage of Ontogeny, human life came to appear entirely dependent on the invariance of classical space, the seamlessness of the continuum. And with the breaching of this continuum in the middle of the nineteenth century, the mirror of ontical identity was broken. It was as if Being looked in the mirror fully expecting to see a being, an object-in-space, as it had for centuries before, and instead "saw its own seeing." Regarding itself in this fashion, Being obtained a backward glimpse of what it really is: the dimension of process that first projects object-in-space-before-subject. Such a Proprioceptive glimmer was hard to take, suggesting as it did the amputation of the old body image. Being had gone so far away from itself in the course of its self-clarification that now, in confronting its own actuality, it saw only death. The wholeness of Being was experienced as but a hole, as an emptiness, as the vacuum left by the radical negation of the most basic ontical expectations. Little wonder that the call to Being went largely unheeded in science, philosophy, and everyday life.

However, given the extraordinary consequences of the call, it was not possible simply to ignore it and go on as before. A response clearly was required. Enter here *modernism*, which may be redefined ontologically as that response to the call to Being that takes Being into account in such a way as to continue denying its true nature. The work of Einstein has been our premier example of modernism in science. In the theory of relativity, entry into the final stage of Ontogeny is forestalled by registering the call to the ontological dimension as a mere extension of ontical dimensionality. Thus, instead of accepting the breach in the three-dimensional continuum as the cue to "reverse gears," to acknowledge the dimension of Being by the Proprioceptive movement backward into it, Einstein maintains the forward, ontical orientation by postulating a *four*-dimensional continuum. A similar stratagem for maintaining spatial invariance was employed by Planck and his successors in their reaction to the peculiar behavior of radiant energy (light) in the microworld. Essentially the same effort to deny discontinuity has been carried forward by those seeking to unify Einsteinian relativity and quantum mechanics. Today, in the multi-dimensional continuum that string theory assumes for its "theory of everything," ontical dimensionality has been stretched to the highest degree of abstraction yet. We discussed all this in Chapter One, and examined the costs of this approach, the price that must be paid for such flights from the "groundless ground" of *apeiron*.

Ultimately, the driving force behind all these efforts at unification is found in the "mirror stage" of ontogeny (Chapter One, pp. 20–21). It was at this early point in development that Being first turned away from itself so as to clarify itself, thereby moving into the second stage of Ontogeny. What this initiated was a quest for unity. Grosz cites Lacan's observation that, in the mirror stage, "'the total form of the body by which the subject anticipates in a mirage the maturation of his powers is given to him only as a *Gestalt*'" (1994, p. 42). Grosz concludes that the "mirror image provides an anticipatory ideal of unity to which the ego will always aspire" (p. 42). However, this "model of bodily integrity" is something "the subject's experience can never confirm....[T]he stability of the unified body image...is always precarious" (p. 43). Nevertheless, just this imagined unity of the ego or subject is,

> the precondition...for all symbolic interactions and for an objective or scientific (i.e., measurable, quantifiable) form of space. The virtual duplication of the subject's body, the creation of a symmetry measured from the mirror plane, is necessary for these more sophisticated, abstract, and derivative notions of spatiality. (Grosz, 1994, p. 45)

So the search for unity (symmetry, continuity, invariance) is what science has always been about. But because the imagined "final unification" is rooted in the unity of an ego that, in fact, is always precarious, complete ontical unity is sought in vain. This was the truth that confronted science in the middle of the nineteenth century, as it reached its culmination. The underlying message was that we could go no further with the ontical project, that Being had given itself all it could in this mode of self-concealment. Therefore, it was now time to *receive* this gift, to turn Proprioceptively backward to Being, thereby making the transition to the next stage of Ontogeny. But modernism turned a deaf ear to this message and the refusal has eventuated in postmodern collapse. In postmodernity, modernism's "phantom-limb experience" loses all credibility and, once again, as in the nineteenth century, Being—still assuming its posture of self-concealment—confronts the amputation of the unitary body.

Ontological phenomenology has sought to provide an alternative to the modernist-postmodern cycle of "bright optimism" and "dark despair." Phenomenological thinking (at least as adumbrated by Heidegger and Merleau-Ponty) attempts to *heed* the call to Being. I venture to say that the onto-phenomenological posture is already in evidence in what I have written heretofore, but much more needs to be done by way of working out the details of the ontological dimension. What sort of work will this be?

The investigation I am referring to is certainly not intended as merely an "objective account" of apeironic Being. The analysis will not unfold within some "meta-space," some higher-order "area of measurement" wherein Being itself is definitively determined for the use of a still more abstract subjectivity, one that is even further removed from the lifeworld of phenomenology. The exploration in question aims at a *reversal* of this ontical gearing; it looks to move backward into the dimension of Being. Therefore, rather than attempting an objectification of Being, the task is to *acknowledge its gift*. Once again, Appropriation is an intimate *giving*, a giving to one of what is one's own. In Being's self-Appropriation, it gives to itself object-in-space-before-subject by way of clarifying itself. The "task of thinking" that still lies ahead is to enter into Being in such a way that its gift to itself is now consciously received. If the process of giving is to be completed in earnest, the gift indeed must be thankfully/thought-fully accepted. In so receiving the gift, we reverse our detached stance and become an integral part of the process that we analyze. This means, of course, that we shall have to do more than approach Being in a conceptual or perceptual way. *An act of Proprioception* surely will be required.

Given that Being is neither an object nor a subject, neither a truth carved in stone nor a flight of fancy, we might say of it what Heisenberg said of the microphysical realm: it constitutes "a strange kind of physical reality just in the middle between possibility and reality" (1958, p. 41). Or, following Bohm's (1951) portrayal of the microphysical wave function, we could say that Being entails "tendencies" or "potentia" (see Stapp, 1986, p. 265). But there is a simpler way to put it. Being is an *organic process*, as already suggested. Therefore, rather than forever remaining the same or changing arbitrarily, it grows and develops. What does this imply for our way of interacting with it?

An objective observer *discovers* the truth, whereas a subjective agent *invents* it, makes it up as he goes along. Exactly what relationship to truth is suggested by the truth of organic process? Can we not maintain an objective posture, despite the fact that we are dealing with that which changes organically? Doesn't modern biology attempt to do precisely that? The problem with this approach is that the internal dynamism of life vanishes the moment we detach ourselves from it, since the old formula then prevails and life process is relegated to the mechanistic transfer of energy in a changeless continuum. Therefore, if we are to relate to the life of Being in a bona fide way, we cannot stand outside of it and describe what is "objectively there"; rather, we must *actively participate* in it. This surely does not mean acting with the free hand of a subject who seeks to mold the living truth in accordance with his own capricious designs (current efforts at genetic engineering notwithstanding). For, if organic Being cannot be rendered changeless without eluding us, neither does this Trickster really lend itself to the arbitrary manipulations of an "unmoved mover," a *deus ex machina* operating upon it from afar. So it seems that neither of the postures of patriarchy—that of passive observer or of active agent—will suffice when it comes to addressing the matter of Being, the life of the *apeiron*. Then what specific attitude toward Being must we assume if we are genuinely to participate in its life process?

A "maternal" attitude, I suggest. This means that, instead of denying Being's dynamic life or seeking to shape it to our will, we need to encourage it, cultivate it, serving as catalysts that enable it to complete its individuation. It is by adopting the Proprioceptive posture that we do this, since the development of the apeironic organism is completed when the gift it has given to itself in self-Appropriation is attentively received. But the "we" who is to engage in this Proprioception surely cannot be a collection of subjects that would be capable of relating to Being only passively or willfully. Since Proprio-ception is a *self*-receiving, the "we" must be Being itself; the transformation brought about by entering the third stage of Ontogeny must be a *self*-transformation. So the Proprioceptive writings of Heidegger and Merleau-Ponty, and our own work here, constitute the self-generative text of *apeiron*. Phenomenological thinking comes into its own, moves into the climactic stage of Ontogeny, when Being assumes the role of author, when the words of the text are Being's own.

In such an organic writing, Being neither recounts "autobiographical fact" nor invents itself fictitiously. It *gives birth to itself.* We have seen that Merleau-Ponty speaks similarly in describing the organic *painting*. I take the liberty of repeating that key passage one more time:

> The painter's vision is not a view upon the *outside*, a merely "physical-optical" relation with the world. The world no longer stands before him through representation; rather, it is the painter to whom the things of the world *give birth* by a sort of concentration or coming-to-itself of the visible. Ultimately the painting relates to nothing at all among experienced things unless it is first of all "autofigurative"....The spectacle is first of all a spectacle of itself before it is a spectacle of something outside of it. (1964, p. 181; italics added)

If, for "painter," we substitute writer, and for "the things of the world," we insert Being (an amendment that Merleau-Ponty would likely have approved), then Being's authorship of this text is confirmed. It is Being itself that gives birth to this writing. It does so by a *self-coagulation*, to substitute the alchemical term for Merleau-Ponty's "concentration or coming-to-itself." Of course, the content of this textual "autofiguration" is not ontical. It is not a fruit bowl, bouquet of flowers or a landscape that Being writes of. As an *ontological* writing, an "ontobiography," Being explicitly writes itself.

The serpentine nonlinearity of the process is highlighted by the fact that the birth of Being comes at the *end* of its development. It is only after all of Being's ontical possibilities have been exhausted that Being can move "backward through the birth canal," return Proprioceptively to the "first possibility for thinking," and bring this embryo to term.

Five

TOPOLOGY

1. Introduction

In Heidegger's adumbration of "time-space," in Merleau-Ponty's intimation of the depth dimension, the Ontogenetic gears are reversed and we are poised to "advance backward" into Being. Now the task at hand is to specify further the ontological dimension in which the ontical world is grounded. Details are required for a full-fledged response to the call to Being. The dimension of Being must be further concretized, fleshed out, made palpably real. For Being to assume full authorship of this text, these still-too-abstract words about Being must be turned into living flesh. The present chapter is dedicated to pursuing this aim.

In a preliminary working note for *The Visible and the Invisible* written in October 1959, Merleau-Ponty instructed himself as follows:

> Take topological space as a model of being. The Euclidean space is the model for [idealized] perspectival being [and is consistent]...with the classical ontology....The topological space, on the contrary, [is] a milieu in which are circumscribed relations of proximity, of envelopment, etc. [and] is the image of a being that...is at the same time older than everything and 'of the first day' (Hegel)....[Topological space] is encountered not only at the level of the physical world, but again it is constitutive of life, and finally it founds the *wild* principle of Logos — — It is this wild or brute being that intervenes at all levels to overcome the problems of the classical ontology. (1968, pp. 210–211)

Unmistakably, this topological space—involving as it does the relations of "proximity" and "envelopment," and being "older than everything and 'of the first day'"—is equivalent to the concept of dimension Merleau-Ponty had outlined earlier (in "Eye and Mind"): the concept of *depth*. In associating Euclidean space with "the classical ontology," Merleau-Ponty in effect was proclaiming it *ontical*. That is because, while classical "ontology" attempts to study the general nature of being, it neglects the "wild side"—the embodied, apeironic dimension that constitutes the *Being* of being. Merleau-Ponty implies that it is *topological* space that is truly ontological.

Note the etymological indication of topology's concreteness compared with other branches of mathematics. Topology is the study of *topos*, "place." The concrete character of this term is evidenced by its relationship to words like "posture": the root meaning of "posture" is "to place." Philosopher of science John Schumacher thus defines "posture" as "the way a thing makes a place in the world" (1989, p. 17–18). Accordingly, the philosopher Maxine Sheets-Johnstone is able to demonstrate that, whereas Euclidean geometry involves practices that are largely disembodied, "topology...is rooted in the body" (1990, p. 42). Is it not the basic re-gearing of bodily posture (concrete orientation) that we require for our ontological reversal? Then topology would seem to be the discipline of choice. This conclusion is supported by the interpretation of topology offered by the communications researcher Anita Hammer. Hammer (1999) examines the relationship between topology and imagination. When a new direction for thinking is indicated, imagination is needed to give it concrete form. And *topological* imagination can further the process by lending precision to this form. Following Hammer's lead, we can say that the topological path on which we now embark is intended to give precise imaginal form to onto-phenomenological thinking so as to flesh it out to the fullest degree. But let me digress briefly to consider the historical background of topology.

There came a time in the history of mathematics when the universality of Euclidean space was challenged. It was the nineteenth century, of course. The Euclidean conception of space is in fact what lay behind the mechanical relations of the Galilean continuum, and the *subversion* of Euclideanism in physics coincided with the breaching of the luminiferous ether that had been posited to maintain Galilean continuity. However, in the field of mathematics proper, the challenge to Euclideanism actually came long before Michelson and Morley's crucial experiment, and the modernist response to this challenge had begun to be formulated at that earlier time.

Early nineteenth century doubts about Euclid's parallel postulate had called into question the classical formula of object-in-space-before-subject by questioning the Euclidean space that had served as the context within which mathematical thinking previously had operated. We may consider this as the opening note in the call to Being. The response of modernist mathematics, however, was to *deny* the call, to uphold the classical projection, maintain the "forward gearing" of consciousness. How did it do this? It rearranged the particular values that were assigned to the ontical formula without changing the formula itself. The modification essentially amounted to transferring the Euclidean continuum, which had served as the "space" term in the old formula, to the "object" term. That is, Euclidean space was no longer the means by which mathematical objects were studied but was now itself an object to be studied in a more abstract spatial context. Within that context, it became possible to investigate mathematical objects of a more general nature. Thus there arose non-Euclidean systems, such as the hyperbolic and elliptic geometries. In the course of the nineteenth century, the Euclidean approach was further surpassed through the development of projective geometry, and beyond that, the highly general study of topology was inaugurated. Therefore—despite the fact that topo-logy, the study of place, of posture, "is rooted in the body" (Sheets-Johnstone, 1990, p. 42)—under the influence of modernism, topology became an instrument for the denial of concrete Being, one that re-imposed ontical continuity at a higher level of abstraction.

Why did Merleau-Ponty look to topological space as a fruitful approach to "wild Being" (*apeiron*)? Apparently he saw beyond the modernist treatment of topology to the root potential of this practice. Merleau-Ponty evidently sensed that, while *modernist* topology can only refuse the call to Being, a phenomenologically grounded topology could offer us something more. I venture to say that, indeed, there are certain curious structures of modern visual geometry and topology that show great promise in this regard. I propose that, if the structures in question can be radically reinterpreted along phenomenological lines, they can help us with the task of further specifying the ontological dimension, concretizing it, fleshing it out with precision. As a matter of fact, Merleau-Ponty himself worked with such a structure.

2. Embodying the Ontological

A. The Necker Cube

In Merleau-Ponty's *Phenomenology of Perception* (1962), he made use of a well-known visual paradox of modern psychology: the *Necker cube* (Fig. 6b, next page). It is this structure, I suggest, that can provide us with a preliminary graphic illustration of the dimension of depth, as is implicit in Merleau-Ponty's own writing. Let us first consider the Cartesian principle of opposition as expressed through visual perspective in Figure 6a.

If you were initially viewing a solid cube from the angle shown in the left-hand member of Figure 6a, you would obtain the point of view of the right-hand member by (1) moving 180° around the cube to the opposite side, and (2) moving above the cube, since the left-hand perspective gives the view from below. The faces of the symbolized solid that are visible from the right-hand perspective are precisely those that were concealed from the left-hand point of view, and vice versa. In our ordinary experience with perspective, it is of course impossible to view both the near side and the far side (or the inside and the outside) of an object simultaneously; all the faces of the

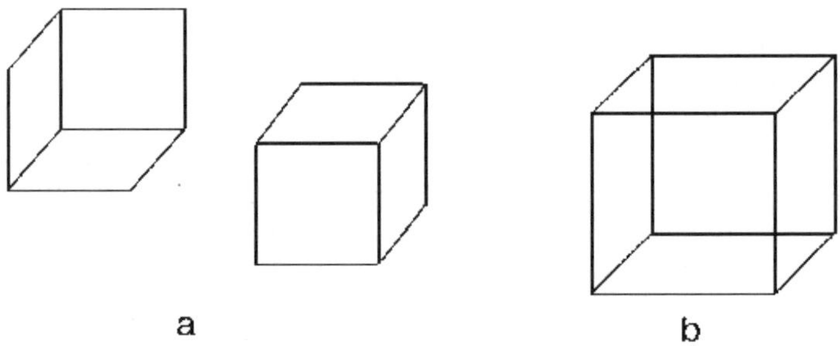

Figure 6. Opposing perspectives (a) and Necker cube (b)

cube cannot be apprehended in the same glance. Opposing faces are closed to each other.

The ordinary mode of perception is the Cartesian one. In this "from-to" posture, we perceive objects and events as extended in the world outside us, but have no immediate access to the inner, subjective ground of our perceptions: we cannot see our own act of seeing, touch our own touching. What Figure 6a illustrates is that this underlying opposition between the subjective seat of perception "in here" and the objective realm "out there" is reflected in the external objects themselves, in the diametrical opposition we ordinarily encounter between their concealed and exposed surfaces. Opposing sides of objects cannot be viewed at once.

Turning now to Figure 6b, inspection readily discloses that both of the perspectives shown in Figure 6a are encompassed in the body of the Necker cube. This creates visual ambiguity. You may be perceiving the cube from the point of view in which it seems to be hovering above your line of vision when suddenly a spontaneous shift occurs and you see it as if it lay below. Two disparate perspectives surely are experienced in the course of gazing at the cube and this disparity reflects the continuing distinction between opposing sides. But the cube's reversing perspectives *overlap* one another in space, are internally related, completely interdependent (think of what would happen to one perspective if the other were erased).

Now, recall Merleau-Ponty's comments on the primal dimension of depth: "[A] *first* dimension that contains all the others is no longer a dimension, at least in the ordinary sense of a *certain relationship* according to which we make measurements. Depth thus understood is, rather, the experience of the reversibility of dimensions" (1964, p. 180). Evidently, the Necker-cube relationship of oscillating perspectives can be taken as illustrative of the reversibility of the depth dimension. In this regard, here is a comment made by Merleau-Ponty himself:

> [I]f the hidden face of the cube radiates forth somewhere as well as does the face I have under my eyes, and coexists with it, and if I who see the cube [I, the "invisible" subject] also belong to the visible, I am visible from elsewhere, and if I and the cube are together caught up in one same "element" (should we say of the seer, or of the visible?), this cohesion, this visibility by principle, prevails over every momentary discordance. (1968, p. 140)

Merleau-Ponty's best-known example of Necker-cube-like reversibility is featured in his last work, *The Visible and the Invisible* (1968). Near the beginning of his pivotal chapter titled "The Intertwining—the Chiasm," he illustrates the interchange of subject and object as a "veritable touching of the touch, when my right hand touches my left hand while it is palpating the things, where the 'touching subject' passes over to the rank of the touched, descends into the things" (pp. 133–134). What we have here is a free reversibility of subject and object wherein, one moment, my left hand plays the role of subject, fingering an object, say, the keyboard of this computer; while, in the next moment, my left hand itself becomes object to the "subjectivity" of my right hand. And this reciprocal relation is not limited to the senses, to touching or seeing. According to Merleau-Ponty, "As there is a reversibility of the seeing and the visible...so also there is a reversibility of speech and what it signifies" (1968, p. 154). This means that the speaking and thinking subject—no less than the sensing subject—is an embodied participant in the earthly transactions of the phenomenological lifeworld, not just a detached *cogito*.

Nevertheless, Merleau-Ponty imposes a significant limitation on the interrelatedness of subject and object:

> [W]e spoke summarily of a reversibility of the seeing and the visible, of the touching and the touched. It is time to emphasize that it is a reversibility always imminent and never realized in fact. My left hand is always on the verge of touching my right hand touching the things, but I never reach coincidence; the coincidence eclipses at the moment of realization, and one of two things always occurs: either my right hand really passes over to the rank of the touched [becomes an object], but then its hold on the world is interrupted [it is no longer a subject]; or it retains its hold on the world [remains a subject], but then I do not really touch *it*—my right hand touching; I palpate with my left hand only its outer covering. (1968, pp. 147–148)

In other words, while what was subject can be known as object a moment later, I cannot know subjectivity as such. The subject is still the one who does the knowing, while the object remains that which is known.

Apparently then, the intimate reciprocity of subject and object, their thoroughgoing entwinement, does not add up to a literal fusion, at least not in terms of what humans can *experience*. To be sure, on Merleau-Ponty's account, subject and object are consanguineously tied to one another; they participate fully in the same order of Being. With little difficulty, subject can become object and vice versa, and this cube-like reversibility attests to the blood-bond inwardly uniting them. Yet "the hinge between them, solid, unshakeable,…[remains] irremediably hidden from me" (1968, p. 148). So, according to Merleau-Ponty, the boundary dividing subject and object holds firm in my experience of them; in passing from one to the other, necessarily, there is a break in my awareness, and this gap attests to their mutual exclusion from each other. Merleau-Ponty does go on to say that "this hiatus between my right hand touched and my right hand touching...is not an ontological void, a non-being: it is spanned by the total being of my body, and by that of the world" (p. 148). Still, while there may not be an ontological void, the presumably "irremediable" *epistemological* void he posits would be quite enough to forever prevent the *realization* of full-fledged subject-object intertwinement.

What are the implications of Merleau-Ponty's self-imposed limitation for the task at hand? We are to heed the call to Being by fleshing out its dimension of depth. To achieve our aim, we must "reverse gears," counteract the prevailing tendency for awareness to move away from *apeiron*. Merleau-Pontean reversibility does not allow us to do this in a full-fledged way, essentially because it remains too abstract. The call to Being does echo in the intimate relatedness of subject and object, but that relationship can only be inferred indirectly from the reversibility Merleau-Ponty observes. In Merleau-Ponty's approach, any particular cognitive or perceptual act is itself actually *ir*reversible, is a uni-directional movement of classical perception from subject to object that maintains the split between them. To repeat the example, on one occasion my left hand serves as subject as I pass from it to the buttons of this keyboard, which are its objects. Then there is the "gap," the "irremediable hiatus," after which it is now my right hand that plays the role of subject, the perspective from which my left hand is known as object. Though the identity of subject and object is thus reversed from the first occasion to the second, what does not change is the irreversibility of the action *within* the given occasion that puts the subject out of reach. In passing to the second occasion, I do come to know what had been the subject, namely, my left hand; but I do not know it *as* subject, since I am presently taking it as my object. So, whereas Merleau-Pontean reversibility changes what is being objectified, it does nothing to directly challenge the act of objectification itself, the act by which we pass irreversibly *from* subject *to* object, thereby occluding the presencing of Being. I am suggesting then that while Merleau-Pontean reversibility connects subject and object in an abstract way, in concrete terms it merely juxtaposes two instances of the *dis*connectedness of the subject and object. As a consequence, we remain unable to bring Being fully to light. I would now like to propose that, beyond the Merleau-Pontean reversal occurring from one occasion to the next, we require a *self-reversal of Being* within the *given* occasion. We require, that is, an act of Proprioception. This can be illustrated by going a step further with the Necker cube.

The limitation of Merleau-Pontean reversibility is symbolically well conveyed in the ordinary way of viewing the cube. Thus—when Merleau-Ponty speaks of the "impotency to superpose exactly upon one another the touching of the things by my right hand and the touching of this same right hand by my left hand," when he insists that "there is always a 'shift,' a 'spread,' between them" (1968, p. 148)—this is depicted quite precisely in the Necker-cube shift in perspective wherein each face that had appeared as an inner surface now presents itself as an outer surface, and vice versa. To be sure, in this shifting of the cube from one perspective to another, we do gain an indirect glimpse of Being; the cube symbolically conveys a general sense of the prereflective dimension of depth from which the reflecting subject and reflected upon object first emerge. Whereas Being operates strictly "behind the scenes" in the external relationship between perspectives pictured in Figure 6a, with the Necker cube of Figure 6b, it becomes more explicit. However, I propose that the cube can be viewed in such a way that the functioning of Being can be made still more explicit.

Let us reconsider the actual manner in which we normally become aware of the cube's reversing perspectives. We are viewing the cube from one point of view when, quite suddenly, we realize that we are now perceiving it from the other. When this happens, we are not actually cognizant of the transition itself, but only of the result of this event. That is, in deducing retrospectively that a shift "must have taken place," we engage in an abstractive reconstruction of Being's concrete act, rather than directly apprehending it. Is there another manner of experiencing the cube? Might there not be a way to depart from perceptual habit and gain a direct glimpse of what lies "between" the opposed perspectives, a concrete insight into the actual event of their reversing?

Merleau-Ponty was not the only student of perception to have categorically denied this possibility ("the hinge between [opposing perspectives remains] irremediably hidden from me"). For example, Asghar Iran-Nejad (1989), after reporting on psychological research said to unequivocally demonstrate the mutually exclusive fashion in which the cube's two perspectives are perceived, buttressed his empirical finding with a logical argument: "To the extent that two different schemas must share [the same] knowledge components, they cannot...be held in mind simultaneously just as two meetings cannot be held at the same time to the extent that the same individuals must participate in both of them" (1989, p. 131). This point of logic boils down to the *principle of non-contradiction*, characterized by philosopher Peter Angeles as one of the basic "laws of thought" in Aristotelian philosophy: "A thing *A* cannot be both *A* and not *A* (at the time that it is *A*)" (1981, p. 153). Angeles further observes that classical logic has been challenged by *dialectical* logic, "a logic of becoming that attempts to present the ever-changing processes of things" (p. 155). Here "contradiction exists in reality. It is possible for the selfsame thing to be and not to be—the same thing is and is not" (p. 155). Since there would be no contradiction in saying that *A* becomes not-*A* with the passage of time, dialectically it is understood that *A* is not-*A* "at the time that it is *A*" (p. 153).

Apeironic Being, as a dialectical life process, certainly does not conform to Aristotelian strictures. We know that Being's temporality is not the mere sequence of point-like, extensionless "nows" ontically posited by Aristotle. It is only if time were Aristotelian that *A* and not-*A* could not be identical to one another in the same instant. But the "true time" adumbrated by Heidegger possesses the dialectical thickness in which the "future, past, and present...belong together in the way they offer themselves to one another" (Heidegger, 1962/1972, p. 14). Thus, in the temporality of Being, it is indeed possible for opposing perspectives of the cube to coincide "simultaneously" (see below for a significant qualification of this term). And it is possible for us to *experience* this concurrence, assuming we do not divide knowing from Being by imposing an artificial restriction on the former. Still, we do have trouble recognizing these possibilities. This is not surprising given the depths of our roots in the classical ground. Understandably, our perception of the Necker cube is strongly influenced by the ontical posture that has long governed human functioning. Therefore, even when we are able to reconstruct abstractly Being's concrete process of reversing perspectives, we find it difficult to grasp that process directly. But this is precisely what we must do if Being is to be embodied more concretely, fleshed out to the fullest extent. And it is something that we actually *can* do.

Instead of allowing our glance to alternate between the opposing perspectives of the Necker cube, we can break this visual habit and view both perspectives at the "same time." I have explored this possibility elsewhere (Rosen, 1986, 1988, 1994, 1997), and it is confirmed in Bruno Ernst's (1986) study of *Belvedere*, a graphic work by M. C. Escher:

Figure 7. M.C. Escher's *Belvedere*

In his analysis of the Escher print, Ernst calls our attention to a detail at the bottom (a little left of center): a boy is sitting on a bench and puzzling over an odd-looking structure he holds in his hands. Lying at his feet is the blueprint for this structure, and, indeed, for the entire artwork. It is the Necker cube. To bring out the underlying principle of *Belvedere*, Ernst provides his own diagram of the cube:

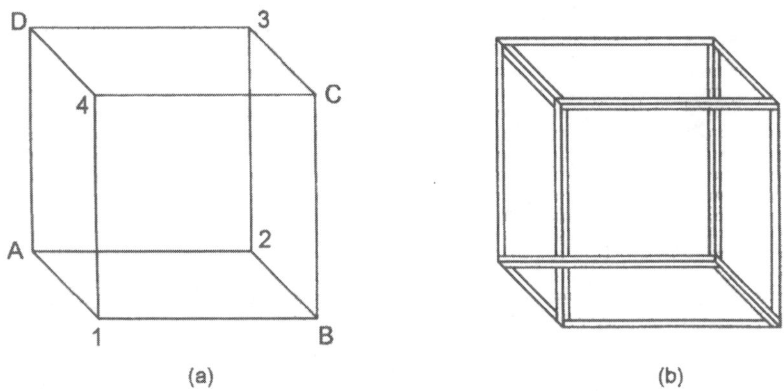

Figure 8. Bare Necker cube (a) and cube with volume (b) (after Ernst, 1986, p. 86)

According to Ernst, the cube encompasses within itself,

> the projection of two different realities. We obtain the first when we assume that points 1 and 4 are close to us and points 2 and 3 are further away; in the other reality, points 2 and 3 are close and 1 and 4 further away....But it is also possible to see points 2 and 4 in the front and 1 and 3 in the back. However, this contradicts our expectation of a cube; for this reason, we do not readily arrive at such an interpretation. Nevertheless, if we give some volume to the skeletal outline of the cube, we can impose said interpretation on the viewer by placing A2 in front of 1–4 and C4 in front of 3–2. Thus we obtain [Figure 8b] and this is the basis for *Belvedere*. (Ernst, 1986, p. 86; translated by M. A. Schiwy)

It is clear that Figure 8b "contradicts our expectation of a cube" because it brings together opposing "realities" (perspectives) that we are accustomed to experiencing just one at a time. When this happens, there is an uncanny sense of self-penetration; the cube appears to do the impossible, to go *through* itself (thus Ernst speaks of constructions based on the cube as "impossible"; 1986, pp. 86–87). If we again imagine the bare cube (Figs. 6b and

8a) as a solid object appearing in space, one whose faces are filled in, we find that perspectival integration has an interesting effect on those faces.

As noted above, in the conventional, perspectivally polarized way of viewing the cube, when the shift is made from one pole to the other, all the faces of the cube that had been seen to lie "inside" presently appear on the "outside," and vice versa. But it is only at "polar extremes" that faces are perceived as *either* inside *or* outside. With the fusion of perspectives that discloses what lies *between* the poles (the "between" that Merleau-Ponty judged "irremediably hidden"), each face presents itself as being inside and outside at the same time. Therefore—at least symbolically—the division of inside and out, of subject and object, is surmounted in the creation of a *one*-sided experiential structure whose opposing perspectives are simultaneously given.

Simultaneously? Well, that is not exactly the case. I am proposing that we can apprehend the cube in such a way that its differing viewpoints overlap in time as well as in space. But what we actually experience when this happens is not simultaneity in the ordinary sense of static juxtaposition. We do not encounter opposing perspectives with the same immediacy as figures appearing side by side in Cartesian space, figures that coexist in an instant of time simply common to them (as, for example, the letters of the words on this page). If the coincidence of the cube's perspectives were limited to that, Merleau-Ponty certainly would be right to say that such opposites could "never reach coincidence." But there is indeed a coincidence in the integrative way of viewing the cube, for perspectives are not related in simple Aristotelian succession (first one, then the other) any more than in Cartesian simultaneity. If opposing faces are not immediately co-present, neither do they disclose themselves merely in *seriatim*, in the externally mediated fashion of linear sequence. Instead the relation is one of internal mediation, of the *mutual permeation* of opposites. Perspectives are grasped as flowing through each other in a manner that blends space and time so completely that they are no longer recognizable in their familiar, categorially dichotomized forms. You can see this most readily in viewing Figure 8b. When you pick up on the odd sense of self-penetration of this "impossible figure," you experience its two modalities neither simply at once, nor one simply followed by the other, as in the ordinary, temporally broken manner of perception; rather, you apprehend an unbroken flowing from one to the other. (Merleau-Ponty himself actually may have been trying to articulate this dialectic of mediation and immediacy when he spoke of "self-mediation" [*médiation par soi*] in "Interrogation and Dialectic," a chapter of *The Visible and the Invisible* that precedes "The Intertwining—the Chiasm." If so, in the relevant passages I have cited from the latter, he does not seem as closely in contact with the dialectic. Here he appears to privilege mediation.)

Its non-simultaneous aspect makes it clear that the perspectival integration of the cube does not just negate the distinction between sides. Faces of the cube *are* inside, and yet outside as well. So the feature of separateness is not lost; rather, a unity is gained that is deeper than that of the simply dipolar Necker cube; without merely eliminating opposition, the complete interpenetration of opposites is embodied. To be sure, this is a paradox, and in expressing it through the concrete act of perception performed with the cube, it is fleshed out more than words alone can do. It thus becomes a lifelike presence, manifesting itself in a way that permits one tangibly to grasp it. Through this body of paradox, one may symbolically gain a palpable glimmer of how subject and object can be opposed, as in reflective consciousness, and also, prereflectively one and the same.

It is this experience that gives flesh to Heidegger's words on "true time" and Merleau-Ponty's writing about spatial depth (despite Merleau-Ponty's own insistence, in "The Intertwining—the Chiasm," on the impossibility of such perception). In the previous chapter we found that, whereas time and space are categorially separated in the Einsteinian light cones, they overlap dialectically in the paradoxical thickening of the here and now intimated by Merleau-Ponty and Heidegger. At that point in the discussion, I suggested that, with the ontological presencing of elsewhere and otherwhen to each other, we might imagine the light cones being retracted, reeled back into the here-now. Such a retraction would amount to a return of radiant energy to its source in Being. I now submit that the perspectival integration of the Necker cube gives concrete expression to this Proprioception of light.

Of course, from the ontical standpoint, the movement of light back into Being would be perceived as a catastrophic rupturing of the luminous continuum to produce a "black hole." Indeed, we know that Einstein's four-dimensional electromagnetic continuum was conceived in the first instance as an attempt to bypass the hole in the three-dimensional luminiferous ether that confronted Michelson and Morley—an attempt that was thwarted in the end by the appearance of black holes in Einstein's new "gravitational ether." Presently it should be clear that what Michelson and Morley encountered is symbolized with graphic precision in the phenomenon of the Necker cube. In both cases there is the "shock of discontinuity" (see Part One) attendant to the fact that old perspectival expectations have been radically upset. The categorical difference in perspective that is classically anticipated is not forthcoming because the common ontological ground that underlies opposing perspectives is now implicitly exposed. When we trace the backward movement of light into the "black hole" via *perspectival integration* of the Necker cube, it turns out that our ontical fears are not confirmed. Instead of encountering the anticipated collapse of being into the singular point at the center of the "hole," we discover that this erstwhile point of departure is the whole dimension of *B*eing.

However, we must now consider an obvious limitation in the exercise with the Necker cube. The cube is a one-dimensional line drawing embedded in two-dimensional space (the space of the printed page). Therefore, while the perception of the cube does simulate our experience with solid entities in three-dimensional space, the cube itself is certainly no solid. I have proposed that the perspectival integration of the cube conveys a sense of the working of Being. But it is clear that such an insight falls short of encompassing Being's activity in the world of three-dimensional objects and events. Therefore, although the non-Aristotelian apprehension of the cube may *symbolize* the transpermeation of subject and object in three dimensions, it is *no more than* symbolic. In the three-dimensional context, when the cube is taken literally rather than figuratively, it is but an object I reflect upon, one that appears before me, a being that is circumscribed, closed into itself, closed off from the inwardness of this subjectivity that does the reflecting. Here the Cartesian *cogito* is not itself opened to view but remains that from which the viewing of object is done. In short, while the one-dimensional cube may effectively *portray* the prereflective activity of three-dimensional Being, in the three-dimensional domain it is actually only a reflected upon being. Is it possible that, if a higher-dimensional version of the Necker cube were available, we would be able to go further?

I suggest the answer is yes. I propose that we would go beyond the merely reflective representation of surpassing reflection's supremacy, that we would make a concrete reality of prereflective Being, if a solid, full-fleshed, three-dimensional counterpart of the cube would stand present. This paradoxical body would stand before me in such a way as also to stand *within* me. It would present itself to me from the inner core of itself and I would recognize that core as my own. That is to say, the three-dimensional object and the dimension constituting my subjectivity would be utterly open to one another, would permeate each other in an unobstructed, boundless exchange. Thus, whereas we do not "lose our objectivity" in perspectivally integrating the one-dimensional cube, whereas the integrated viewing of this object does not encompass our own subjective viewing process, the perspectival integration of a three-dimensional body would be different. Now integration of what is "out there" would be achieved through a process that would carry us back to what lies "in here," to our ownmost lived subjectivity, that is—to apeironic Being itself.

B. Topological Bodies of Paradox

In preparing the approach to a thoroughgoing embodiment of Being, the Necker cube experience of perspectival integration may be expressed in a more tangible form. For this we follow Merleau-Ponty's lead (1968, pp. 210–211) and turn to topology, where we begin with a comparison (Rosen, 1975a, 1986, 1994).

A cylindrical ring (Fig. 9a) is constructed by cutting out a narrow strip of paper and joining the ends. The surface of Moebius (Fig. 9b) is produced simply by giving one end of such a strip a half twist (through an angle of 180°) before linking it with the other.

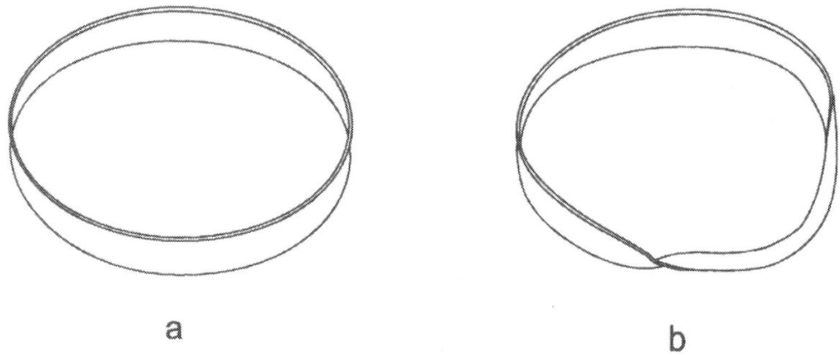

Figure 9. Cylindrical ring (a) and Moebius strip (b)

The cylindrical ring possesses the conventionally expected property of two-sidedness: at any point along its surface, two distinct sides can be identified. Now, in the Moebius case, it is true that if you place your index finger anywhere on the surface, you will be able to put your thumb on a corresponding point on the opposite side. The Moebius strip does have two sides, like the cylinder. But this only holds for the local cross-section of the strip defined by thumb and forefinger. Taking the full length of the strip into account, we discover that points on opposite sides are intimately connected—they can be thought of as twisting or dissolving into each other, as being bound up internally. Accordingly, mathematicians define such pairs of points as *single* points, and the two sides of the Moebius strip as but *one* side. (If the Moebius property of one-sidedness is difficult to imagine in the abstract, it is very easy to demonstrate. For instance, when you draw a continuous line along the whole length of the strip, upon returning to your point of departure you will discover that your ink mark has covered *both* sides of the surface.)

It is important to recognize that the surface of Moebius is not one-sided in the homogeneous sense of a single side of the cylindrical ring. It is one-sided in the paradoxical sense, one-sided and also two-sided, for the local distinction between sides is not simply negated with expansion to the Moebius as a whole. In coming to interpenetrate each other, the sides do not merely lose their distinct identities. Moebius oneness is essentially similar to the oneness of the perspectivally fused Necker cube. There is inside and there is outside. The two are different. Yet they also are one and the same.

The relationship between the Moebius surface and the Necker cube can be understood as analogous to that between a sculpture and a painting (respectively). The two art forms are both external representations of inner dimensions of experience (thoughts, intuitions, feelings). But the sculpture, by making significant use of three dimensions instead of being limited to two, can express the subject-matter more concretely, flesh it out through the tactile sense as well as the visual. In like manner, since the Moebius strip is a two-dimensional surface embedded in three-dimensional space, it can embody the paradoxical union of opposites more concretely than can the lines of the schematic cube, limited as they are to a two-dimensional medium of expression.

Nevertheless, while the Moebius model manifests one-sidedness more tangibly than the cube, it *is* a model, an outward symbolization of the union of inside and out, rather than a full-fledged embodiment directly incorporating the inner depths of Being. What would be needed for the latter? Not a two-dimensional body enclosed as mere object in three-dimensional space, but a body of paradox that is itself three-dimensional.

Let us return for a moment to the one-dimensional Necker cube. In viewing its constituent perspectives as they are separately displayed in Figure 6a, you already may have noticed that each of these components of the reversible cube is itself reversible: you can view either perspective in either a concave or convex fashion. Moreover, from the Necker cube fusion of the opposing perspectives of Figure 6a given in Figure 6b, a new order of *opposition* arises:

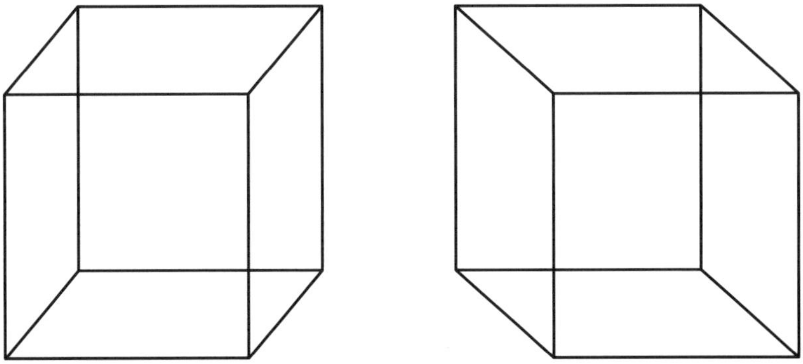

Figure 10. Necker cube opposites

Figure 10 shows that the Necker cube itself possesses a mirror opposite with which it presumably could be integrated to create a third order of self-reversal. There is an unmistakable dialectical pattern at play here wherein the same action that resolves mirror opposition at one level creates new opposition at another. We may study this pattern more effectively in the denser, more concrete medium of the two-dimensional surface.

Mathematicians have investigated the transformations that result from bisecting topological surfaces. Suppose we were to cut the two-sided cylindrical ring of Figure 9a down the middle, proceeding along its full length. Upon completing the cut, the ring would simply decompose into a pair of identical narrower rings each possessing the same topological structure as the original. A more interesting result is obtained in bisecting the one-sided Moebius strip. Instead of falling into two separate pieces as one might expect, the bisected surface retains its integrity but has now become the *two*-sided structure depicted in Figure 11.

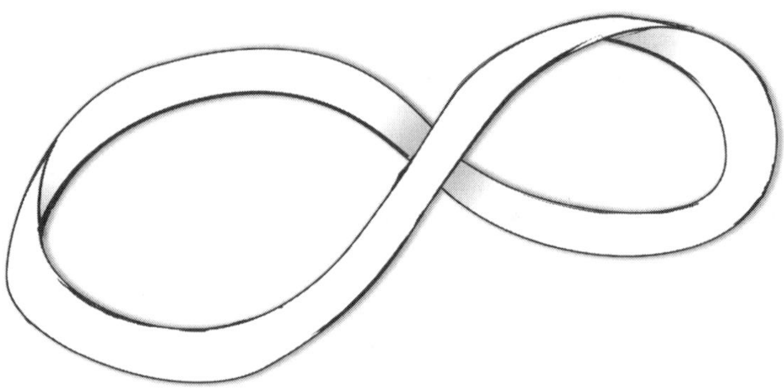

Figure 11. The lemniscate

Compare a single side of this two-sided surface with that of the two-sided cylindrical ring. Whereas revolution about the latter describes but one closed loop, traversal of the former gives us a doubly-looped, figure-8 pattern known in mathematics as a *lemniscate*; turned on its side, the lemniscate is the familiar sign for infinity, ∞. The two patterns of movement are schematically contrasted in Figure 12.

Now, we have found that the integrative quality of the Moebius surface lies in its paradoxical one-sidedness. Like the internally reversing perspectives of the Necker cube, the two sides of the Moebius flow unbrokenly into each other to form a single side, without either side actually losing its distinctness. When bisection of the Moebius strip transforms it into a *two*-sided structure, this integrity is lost. Yet we can now see that each of those sides, being lemniscatory in character, constitutes its own order of paradoxical

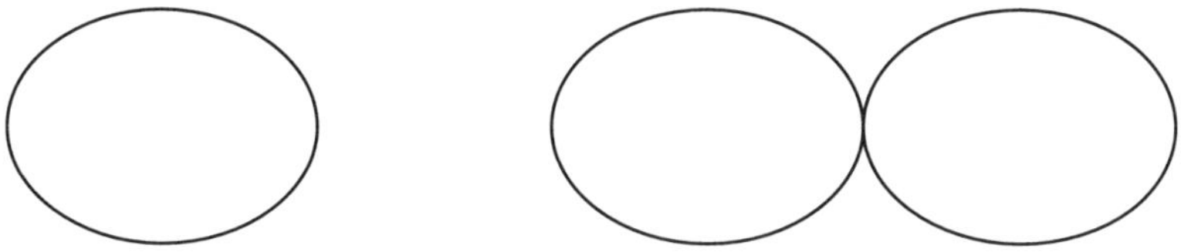

Figure 12. Schematic comparison of cylindrical (left) and lemniscatory (right) rotation

"unity-in-diversity." That is, in a single side, we have a double cycle, these cycles being connected by a continuous movement through the central node of the infinity figure, ∞. It is true that, in the lemniscate, we no longer have a complete overlapping of opposing elements, as we do in the Moebius. The lemniscate thus could be said to have less internal coherence than the Moebius. Be that as it may, a similar pattern of "transpolar flow" is evident in both structures.

The relationship between the lemniscate and Moebius surface is clearly analogous to that between the first- and second-order cubes shown in Figure 6. Just as each of the opposing cubes of Figure 6a constitutes an internal self-reversal in its own right, each of the opposing sides of the bisected Moebius strip illustrated in Figure 11 is composed of the integrated cycles of the lemniscate. And just as fusion of the externally related cubes of Figure 6a gives the second order of cubic self-reversal shown in Figure 6b, fusion of the opposing lemniscates (by gluing together the edges of the bisected Moebius along the seam to undo the bisection) restores the one-sided Moebius in its original integrity. Furthermore, similar to the way a still higher order of integral self-reversal is intimated in the dialectic of the cubes (see Fig. 10), a new order of integration is indicated for the paradoxical surfaces.

Like the cubes of Figures 6 and 10, our surfaces are *enantiomorphic* in nature; they possess the quality of asymmetry entailing mirror opposition. Thus, the single lemniscate has its mirror counterpart on the other side of the bisected Moebius strip, and the Moebius as a whole has its own enantiomorph. Whereas there exists but one form of the symmetric cylindrical ring, the Moebius surface can be produced in either a left- or right-handed version (depending on the direction in which it is twisted). If both a left- and right-oriented Moebius strip were constructed and then glued together, superimposed on one another point for point, a topological structure called a *Klein bottle* would result (named after the German mathematician, Felix Klein).

The Klein bottle (Fig. 13) has the same property of asymmetric one-sidedness as the two-dimensional Moebius surface, but implicates an added dimension (Rosen, 1975a, 1975b, 1980, 1994). Mathematicians tell us that

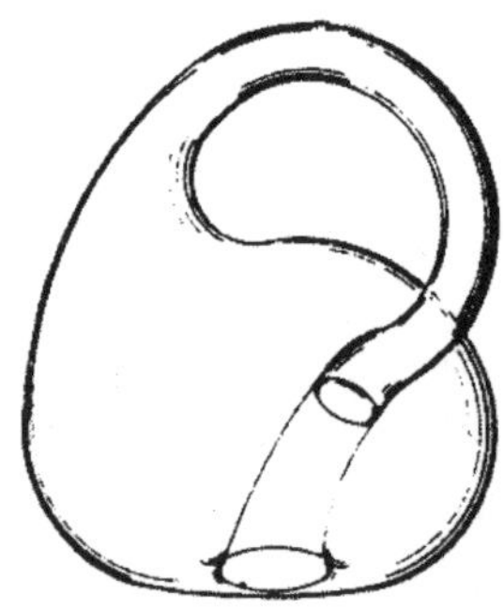

Figure 13. The Klein bottle (from Gardner, 1979, p. 151)

we cannot really produce a proper physical model of this curious bottle. That is, left- and right-facing Moebius bands cannot be superimposed on each other in three-dimensional space without tearing the surfaces. I am going to suggest that this inability to objectify the Klein bottle in three-dimensional Cartesian space actually derives from the fact that the bottle calls into play the dimension of *apeiron*!

There is a different but mathematically equivalent way to describe the making of a Klein bottle that, for our purposes, will be very instructive. Once again a comparison is called for. Both rows of Figure 14 depict the progressive closing of a tubular surface that initially is open. In the upper row, the end circles of the tube are joined in the conventional way, brought

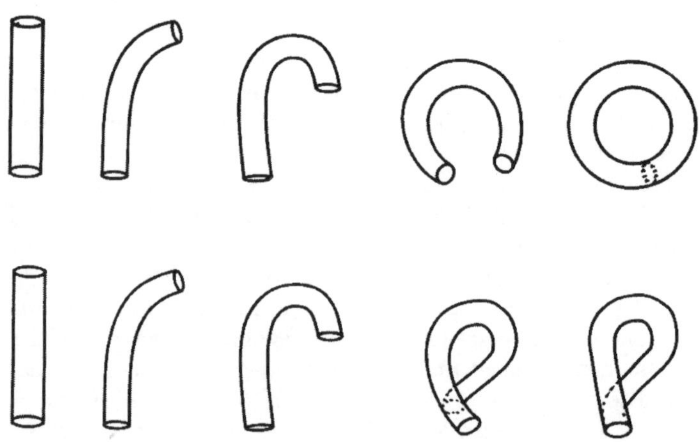

Figure 14. Construction of torus (upper row) and Klein bottle (lower row)

together through the three-dimensional space outside the body of the tube to produce a doughnut-shaped form technically known as a *torus* (a higher-order analogue of the cylindrical ring). By contrast, the end circles in the lower row are superimposed from *inside* the body of the tube, an operation requiring the tube to pass *through* itself. This results in the formation of the Klein bottle. Indeed, if the structure so produced were cut in half, the halves would be Moebius bands of opposite handedness. But in three-dimensional space, no structure can penetrate itself without cutting a hole in its surface, an act that would render the model topologically imperfect. So, from a second standpoint, we see that the construction of a Klein bottle cannot effectively be carried out when one is limited to the three Cartesian dimensions that frame our experience of external, objective reality.

Mathematicians are aware that a form that penetrates itself in a given number of dimensions can be produced without cutting a hole if an *added* dimension is available. The point is nicely illustrated by mathematician Rudolf Rucker (1977). He asks us to imagine a species of "flatlanders" attempting to assemble a Moebius strip. Rucker shows that, since the "physical" (externally experienced) reality of these creatures would be limited to *two* dimensions, when they would try to make an actual model of the Moebius, they would be forced to cut a hole in it. Of course, no such problem arises for us human beings, who have full access to three external dimensions. It is the making of the Klein bottle that is problematic for us, requiring as it would a fourth dimension. Try as we might, we find no fourth dimension "out there" in which to execute this operation.

I venture to propose that the "fourth dimension" needed to complete the formation of the Klein bottle engages the dimension of Being. It is not just another arena for reflection, an "area of measurement" (Heidegger, 1962/1972, p. 15) that stretches before us; rather, it is folded within us, entailing the prereflective depths of *apeiron*. But to fully grasp the nature of the Klein bottle's "missing dimension," it will help to better understand the meaning of the *number* that is assigned to a dimension. A digression is therefore in order.

Classical space is assumed to be three-dimensional. According to Kant, the three-dimensionality of space is not arbitrary but given to us *a priori*; the space of human experience could not be anything other than fully three-dimensional. But it was not long after Kant's death that the "shock of discontinuity" exposed the *ideality* of the classical continuum. We have seen how the Michelson-Morley experiment wound up invalidating the notion that three-dimensional space plus time constitute the seamless universal framework for the objective determination of all that there is. What this crucial experiment intimated is that space is in fact incomplete, that the classical appearance of unbroken continuity conceals a gap. With regard to the question of dimension number, the experiment may be interpreted as suggesting that what is projected as wholly three-dimensional is actually something less. To be sure, the space of human experience is not just *two*-dimensional; but instead of being fully three-dimensional as the idealization implies, it is only "fractionally" so.

The actual incompleteness of classical space seems to be indicated by certain concrete facts of perspectival experience. Let us say there is a box sitting on a table. The number of surfaces of the box that are simultaneously perceptible to me reflects the dimensional scope of my perception. Were my perceptual repertoire strictly limited to two dimensions, I would be able to view but one surface of the box at a time. My experience would thus be *non*-perspectival; it would lack depth, being restricted to the perception of a single plane; I would have no access whatever to the dimension perpendicular to that plane. This is clearly not the case for the human experience of perspective. But neither can we directly perceive all six surfaces of the box at once, as we could if perception were fully three-dimensional. We can simultaneously view all four lines bounding a two-dimensional square; we can perceive at once both of the end-points that bound the one-dimensional line; yet when it comes to the six surfaces that bound the cubic structure, we are limited to viewing a maximum of *three* in the same glance.

Yet what *assumption* do we normally make about this state of affairs? We hardly assume that the surfaces of the box that are hidden from us are simply not there. Since the mirror stage of early childhood, since the emergence of object constancy from the infantile flux, the assumption has been that the whole object is "out there in space," though I cannot *see* all of its faces at any given moment. From this it is clear that the idealization of the object-in-three-dimensional-space is no recent innovation. The Kantian continuum is rooted in the very first projections of object-in-space. Critical here is the separation of what I actually see from what I *think*, with the latter gaining ascendancy. Though I only *see* three faces of the box, I "*know* better," and this alleged knowledge structures my experience in a decisive way.

But this was the knowledge that was called into question after the time of Kant. Never before had the incompleteness of three-dimensional space risen so close to the surface of awareness; in the interest of maintaining constancy (invariance, symmetry), it had always been occluded. The completeness of space, of the objects therein, and of the detached subject who observes them, are entirely interdependent, of course. With the very integrity of the *cogito* thus seeming to hinge on maintaining the projection of this completeness, it is not surprising that—rather than face the actual fractionality of object-in-space-before-subject that was disclosed at mid-nineteenth century—we boarded the modernist-postmodern roller coaster ride that has constituted the crisis of our culture ever since.

In the "up phase" of the cycle, the less-than-three-dimensional character of space is denied even more emphatically than before. The mood here is manic, "hyper" or "hot" (to use McLuhan's term for modernism). The posture of object-in-space-before-subject is maintained despite the *hole* in space, by expansively invoking a "hyper-dimensional" spatial continuum, as in Einstein's response to the Michelson-Morley experiment and in the Hilbert space of quantum mechanics. But then there are the "low points," the mood of deflation and despair found in postmodernism, where it is realized (if only semi-consciously) that the hole *cannot* effectively be denied by modernist extrapolations of continuity. This is the depressive or "cool" phase in the cycle. We want to avoid a regression to infantile chaos but those "cold black holes" tell us we cannot. The underlying fear is of lapsing back into the *pre*-perspectival situation, having our individuality smothered in the darkly oppressive environment of the maternal womb. Haunted by nightmares of dissolution, the call back to the apeironic Mother—Being's call to acknowledge Proprioceptively the gift she has given to herself through her "child"—is scarcely anything the "child" feels like heeding. It is the "child's" worst nightmare because, in this call, it hears only in-completeness and discontinuity; only its own negation.

Nevertheless, the dialectic operating here is such that it is precisely by heeding the call to recognize and enter into the incompleteness of three-dimensional space—the "*child's*" incompleteness, its continuing dependence on the "mother," the subject's lack of simple and total autonomy—that completion is realized in earnest. By *recognizing* the incompleteness, by bringing *conscious awareness* to the "black hole," by going beyond postmodernism and *concretely engaging* the "blackness"—the darkness is lit and what was only fractionally three-dimensional now becomes fully three-dimensional.

Yet the fulfillment attained is indeed dialectical. Consequently, three-dimensional space would not reach completion *as* a continuum, but as a paradoxical circulation of continuity and discontinuity. After all, it is not the *subject's* development that is primarily at stake but that of Being. Therefore, development does not culminate in accordance with the one-sided phallic fantasies of the subject—"total autonomy," "complete transcendence," "pure space," "pure light"—but with a fullness that is at once an emptiness, a wholeness that paradoxically maintains the hole.

We may think the (w)holeness of Being as the dialectic of light. In the second stage of Ontogeny, the shadows are concealed and light is idealized. Thus projected is the three-dimensional continuum, the field of unremitting brightness, of total exposure: that "space without hiding places" (Merleau-Ponty, 1964, p. 173), that purely "positive being" (p. 174), that sheer externality described by Heidegger as the "'outside-of-one-another' of the multiplicity of points" (1927/1962, p. 481). But with the coming of the Age of Light, we are called on to turn our attention from the beings that are lit up in space to light itself, whereupon we discover that the luminiferous continuum is not so continuous as we had supposed: there are "black holes" in it. Superseding modernist/postmodern reactivity, heeding the call to Being, means moving backward into those "holes" in recognition that light as such is not pure positivity (phallic fantasies to the contrary notwithstanding) but is intimately blended with shadow (see Glen Mazis's [1987] phenomenological interpretation of the mid-nineteenth century paintings of J. M. W. Turner). With this act of Proprioception, Being acknowledges the gift it has given to itself, the self-Appropriation that first made possible the bright projection of Kantian space. In this dialectical way, three-dimensional Being enters the climactic stage of Ontogeny, achieves its (w)holeness.

But it is clear that these abstract words about (w)holeness do not in themselves give us access to the concrete reality. In passing from *solve* to *coagula*, our rarefied reflection on (w)holeness must be transmuted into prereflective flesh. It is here that the Klein bottle comes into play. For I have proposed and will attempt to demonstrate that the "higher dimension" needed to complete the formation of this embodied reflective structure is none other than the dimension of prereflective Being. However, the standard mathematical approach to the Klein bottle is a modernist one. In accordance with the ontical formula of object-in-space-before-subject, the Klein bottle is treated strictly as a mathematical *object.*

In conventional topology, the separation of the mathematical object from its containing space is reinforced by dividing container and contained along dimensional lines. To see how this is done, consider again the example of the box. If the experience of a box sitting on a table were strictly two-dimensional, we would be limited to perceiving but a single face of it at a time. The fact of perspective is that we can view the angular relationships of as many as three faces of the box at once. If we were viewing a ball instead of a box, the angularity would be expressed as continuous curvature in the direction perpendicular to the plane of perception. Our ability to perceive angle and curvature attests to the practical truth that the perspectival experience of these objects actually exceeds strict two-dimensionality. How is this handled with the mathematical objects of modern topology? The "objects" we have been working with are conventionally classified as *surfaces*. The lemniscate, torus, cylindrical ring, and Moebius strip are considered to be two-dimensional surfaces embedded in three-dimensional space. But these structures are not simply flat, are they? Does not their curvature in the direction perpendicular to the plane demonstrate that their dimensionality is actually greater than two? The answer given is that the curvature of these topological surfaces is a "global" feature of them that results from the way they are *embedded* in three-dimensional space; this curvature does not alter the intrinsic dimensionality of the surfaces; *as* surfaces, they remain two-dimensional. Similarly, the Klein bottle is taken as a two-dimensional surface that would require a four-dimensional space for its containment. In this modernist analysis, the "fourth dimension" needed to complete the formation of the Klein bottle remains an extensive continuum, though this "higher space" is viewed as "imaginary"; the Klein bottle, for its part, is regarded as an "imaginary object" embedded in this space (whereas the cylindrical ring, lemniscate, Moebius surface, and torus are "real" mathematical objects in the sense that tangibly perceptible models of them may successfully be fashioned in three dimensions). Whether the mathematical object must be approached through hyper-dimensional abstraction or it is concretizable, the mathematician's attention is always directed outward toward an object, toward that which is cast before his or her subjectivity. To repeat, subjectivity itself is the detached position *from* which all objects are viewed, or better perhaps, from which all is viewed as object. Never is subjectivity as such opened to view. In this way, the classical split between object and subject is upheld and the rule of the disembodied subject prevails.

Now, in his phenomenological study of topology, the mathematician Stephen Barr advised that we should not be intimidated by the "higher mathematician....We must not be put off because he is interested only in the higher abstractions: we have an equal right to be interested in the tangible" (1964, p. 20). The tangible fact about the Klein bottle that is glossed over in the higher abstractions of modernist mathematics is its *hole*. Because the standard approach has always presupposed extensive continuity, it cannot come to terms with the inherent *dis*continuity of the Klein bottle created by its self-intersection. Therefore, all too quickly, "higher" mathematics circumvents this concrete hole by an act of abstraction in which the Klein bottle is treated as a properly closed object embedded in a hyper-dimensional continuum. Also implicit in this modernist approach is the detached subjectivity of the mathematician before whom the object is cast. I suggest that, by staying *with* the hole, "lingering awhile" with it, we may bring into question the ontical conception of object-in-space-before-subject.

Let us look more closely at the hole in the Klein bottle. This loss in continuity is *necessary*. One certainly could make a hole in the torus, or in any other object in three-dimensional space, but such discontinuities would not be necessary inasmuch as these objects could be properly assembled in space *without* rupturing them. It is clear that whether an object like the torus is cut open or left intact, the closure of the space containing that object will not be brought into question; in rendering such an object discontinuous, we do not affect the assumption that the space in which it is embedded is simply continuous. With the Klein bottle it is different. Its discontinuity does speak to the supposed continuity of three-dimensional space itself, for the necessity of the hole in the bottle indicates that space is unable to contain the bottle the way ordinary objects appear containable. Topologists tell us that if the Kleinian "object" is properly to be closed, assembled *without* a hole, an "added dimension" is required. Thus, for the Klein bottle to be accommodated, the idealized three-dimensional continuum must in some way be opened up, its continuity opened to challenge. Of course, we could attempt to sidestep the challenge, to skip over the hole by a continuity-maintaining act of abstraction, as in the standard mathematical analysis of the Klein bottle. Assuming we do not employ this stratagem, what conclusion are we led to regarding the "higher" dimension that is required for the completion of the Klein bottle? If it is not an extensive continuum, what sort of dimension is it?

My proposition is that the Klein bottle's "missing dimension" engages the *ontological* dimension of human being; that it is not just another framework for reflection but a dimension that entails the prereflective depths of Being. In realizing Kleinian dimensionality, we do not add a fourth dimension to a three-dimensional space that already is simply complete. Instead, we disclose the actual incompleteness of three-dimensional space, its underlying fractionality or hole. Then, by "reversing gears," by moving consciously backward into the hole, the light of awareness filters back, blending with the darkness. Three-dimensionality is thus brought to dialectical fruition, rendered (w)hole.

Let me now attempt to make this clearer by refocusing on the threefold categorial disjunction implicit in the standard treatment of the Klein bottle: contained object, containing space, uncontained subject. To reiterate: (1) The contained constitutes the category of the bounded or finite, that of the immanent contents we reflect upon, whatever they may be. These include empirical facts and their generalizations (which may be given in the form of equations, invariances, or symmetries). (2) The containing space is the contextual bounded*ness* serving as the means by which reflection occurs. (3) The uncontained or unbounded is the transcendent agent of reflection, namely, the subject. It is in adhering to this ontical trichotomy that the Klein bottle is conventionally deemed a "two-dimensional object" (a surface) embedded in "four-dimensional space." But the actual nature of the Klein bottle suggests otherwise. The concrete necessity of its hole indicates that, in reality, this bottle is not a mere object, not simply enclosed in a continuum as can be assumed of ordinary objects, and not opened to the view of a subject that itself is detached, unviewed (uncontained). Instead of being contained in space, the Klein bottle may be said to *contain itself*, thereby superseding the dichotomy of container and contained. Instead of being reflected upon by a subject that itself remains out of reach, the self-containing Kleinian object may be said to *flow back into* the subject thereby disclosing—not a detached *cogito*, but the presencing of Being.

With respect to dimension number, the Klein bottle's ontological blending of the contained object and its containing space confounds the idealized division of these terms into categorically distinct integer dimensions. The dimensional gap imposed by modernist mathematics is thus closed: the Kleinian object and its spatial context are of the *same* dimension. This surely does not mean that the Klein bottle is *two*-dimensional, as in its topological idealization. But neither is it simply three-dimensional, as is assumed of the classical object embedded in the Cartesian continuum. Instead, the Klein bottle uniquely accords with the phenomenological fact that human experience is "fractionally less" than three-dimensional. Then are we to think of the Klein bottle's dimension number as some definite fraction lying on a scale between the numbers two and three? To think that would be to presuppose ontical continuity. The Klein bottle's dimension number is not "less than three" in the sense of constituting a finite value on a fixed numerical continuum (say, "2.5"). Kleinian dimensionality must be taken as "less than three" in a *qualitatively* negative sense: it possesses a hole that renders the three-dimensional continuum incomplete. The "fractionality" of Kleinian dimension thus points to radical discontinuity, not to a less than three-dimensional *continuum*. Of course, Kleinian dimensionality is not *merely* discontinuous, since Proprioceptive attention to its hole brings dialectical (w)holeness. What we have then in the Klein bottle is the paradoxical completion of three-dimensionality in which incompleteness or discontinuity is not simply eliminated.

The "fractional integrality" of the Kleinian dimension is no doubt a difficult notion to grasp and I will attempt to elucidate it via a schema for the Klein bottle provided by the communications theorist Paul Ryan (1993, p. 98). My adaptation of Ryan's diagram is given in Figure 15 (see next page).

Ryan identifies the three basic features of the Klein bottle as "part contained," "part uncontained," and "part containing." Here we see how the part contained opens out (at the bottom of the figure) to form the perimeter of the container, and how this, in turn, passes over into the uncontained aspect (in the upper portion of Fig. 15). The three parts of this structure thus flow into one another in a continuous movement that flies in the face of the classico-modernist trichotomy. But we can also see the aspect of *dis*continuity in the area of the diagram where a self-intersection is required (the region marked off by the rectangular box). Therefore, in its highly schematic way, the one-dimensional diagram lays out symbolically the basic terms involved in the "continuously discontinuous" circulation of dialectical Being. Is the resemblance between the self-containing Kleinian dimension and Merleau-Ponty's

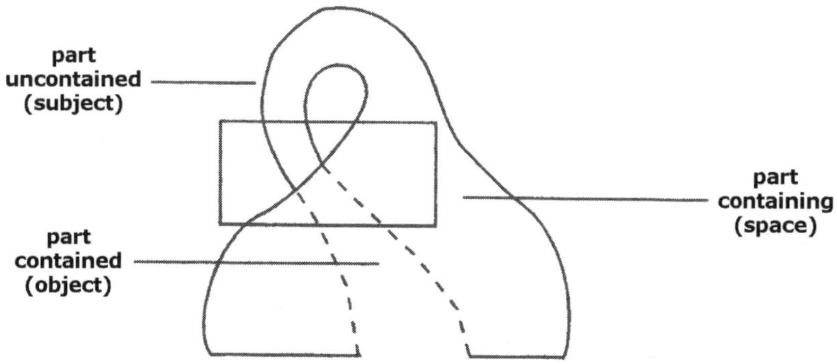

Figure 15. Parts of the Klein bottle (after Ryan, 1993, p. 98)

depth dimension not unmistakable? Dare I say that the former provides the topological detail necessary to flesh out the latter?

Perhaps you have noticed another resemblance. Toward the end of Chapter Three, I addressed the great challenge of "swallowing the serpent," that is, of accepting *apeiron*. In alchemical terms, this is the challenge of containing the tempestuous Spirit Mercurius. S/he must be contained *without* being contained, since we must recognize that—as the uroboros—s/he contains herself (see Fig. 5). For this a special containment vessel is required, one whose contents are hermetically enclosed by *allowing them to spill out*! In my 1995 essay "Pouring Old Wine into a New Vessel," I established that the uroboric vessel of old bears a remarkable affinity to the Klein bottle (compare Fig. 5 with Figs. 13–15). The upshot of the correspondence for our present purpose is this: To contain itself, the serpent must intersect itself, an operation requiring a hole (corresponding to the opening that is its mouth). The hole in the Klein bottle is of just this sort. It is neither solely a hole in a container, nor a hole in that which it contains, but the hole produced by the act of *self*-containment that dialectically integrates the container with its contents in this way giving (w)holeness.

The process that took place in the alchemical vessel was often described as a "circular distillation" (Jung, 1970, p. 11), a transformative circulation of substances "that was in some way turned back upon itself" (Jung, 1967, p. 148). It is the "gear-reversed" Kleinian circulation—from contained object to containing space back to uncontained subject—that fluidly unites the reflective and prereflective in a concretely self-reflective embodiment of three-dimensional Being. The uroboric Klein bottle is a reflected-upon content that, in containing itself, flows unbrokenly backward into its own prereflective ground. That ground is the Klein bottle's "missing dimension." We may indeed say that *all* reflected upon contents originate in the prereflective. But in the case of an ordinary content, we cannot move back into its ground without obstruction because this content lends itself to the appearance of being simply contained, closed into its spatial container in such a way that it is closed off from its prereflective source. Only a *self*-containing object of reflection can incorporate its prereflective origin without a break. To be sure, reflection continues in this self-reflective act. Like the movement from one side of the Moebius strip to the other that paradoxically keeps us on the same side, the retrograde Kleinian movement from reflection to prereflectivity at once maintains the reflective. But now, instead of one-sided domination by reflection, there is harmony with the prereflective.

C. Making Being a Concrete Reality

I hasten to admit that the foregoing words about the Kleinian embodiment of Being may themselves seem rather *dis*embodied. What can we do to make them more concrete?

As a first step, let me acknowledge that, until now, I implicitly have been assuming the very ontical posture I have sought to question. As stressed above, in the ontical situation of object-before-subject, only the object is open to view; the subject remains detached, out of reach, anonymous. And this is essentially how I have approached the "object" of primary concern in this book. In largely maintaining my own anonymity, in failing fully to situate *myself* in this text as the subject before whom apeironic Being is "cast," I have let Being appear as a free-floating abstraction. Then it seems that if I am truly to challenge the ontical stance, I must make Being less remote by recognizing in explicit terms that, rather than standing before some anonymous subject, it stands before *this* subject. That is to say, Being is the content of *this text*; it appears before me, the one who writes these words, and before you who reads them. Surely I state the obvious, but it is precisely the obvious that we lose sight of when we are lost in abstraction. Thus brought down to earth, the question of realizing Being, of embodying three-dimensional (w)holeness through the Klein bottle, becomes one of whether this text we are working with, being Kleinian in character, can lead us back into the prereflective ground from which this reflection of ours originates, and can do so without a break, thereby surmounting the exclusive rule of reflection.

The classical text operates squarely within the reflective mode and raises no questions about doing so. Here the word or sign, whose signifier serves as surrogate for the subject, refers solely to what is other, making this signified object of reflection explicit, a well-bounded content closed into its context. The signifier/subject per se remains implicit; it does not meaningfully refer to itself. Is classical reflection effectively challenged in modernist or postmodern deconstructionist writing? In preceding chapters, there have been ample indications that it is not.

It is true that the modernist sign is self-referential. In modernism, attention is withdrawn from the end-products of reflection and meaning is relocated in an abstraction of the process itself. The sign is thus turned back upon itself so as to bring to light what formerly had been strictly implicit. Or, to put it as I have before, modernism wants to surpass classical signification by turning the classical subject into an object. What actually happens is that the classical posture is maintained at a higher level of abstraction. For, in making the old subject explicit, in rendering it a well-delineated content enclosed in its context, this object-*née*-subject implicitly must be given to, must appear before, a new, higher order subject. That is the crux of what we found in examining modernist science and philosophy, modernist art and media. The self-reflection of modernism is indeed akin to gazing at the reflection of one's eyes in a mirror: what had been the *gaze* of the subject now itself appears as an object gazed upon by a subject that is one step removed from the original. The important thing to keep in mind is that this transformation of terms leaves completely intact the classical relationship of object-in-space-before-subject. At bottom then, the move to modernism poses no fundamental challenge to the mode of signification that preceded it. Modernism's self-referential sign, by turning the self into an other, in fact preserves in abstraction the classical *split* between self and other, between the signifier and its signified object.

In postmodernism, the ultimate consequence of modernism is recognized and played out. We have seen that postmodernism basically entails the realization that modernism's objectification of the subject is but the first term of an infinite regress. In modernism, the classical subject loses its privileged position as the unquestionable base of knowledge, as the transcendent, never-to-be-viewed perspective point from which all else is viewed. Once this subject is viewed, made explicit, objectified, cast before the perspective point of a newly implied, higher order subject, *no* subject can securely hold its position. Having established that the classical subject can be turned into an object, the new, modernist subject should be susceptible to the same fate. The objectification of that subject would bring a still newer order of subjectivity with the same susceptibility, and so on, *ad infinitum*. And each time the subject is undermined by being made into an object, what had been object to that subject is also undermined. Ultimately then, we have neither subject nor object in any stable, abidingly meaningful form.

In the parlance of postmodern or poststructuralist literary theory, the fixed relationships between particular signifiers and their signified meanings give way to the restless, ever-shifting text, the ceaselessly self-alienating application of the sign to itself. Thus Spivak, echoing the outlook of Derrida, says that "[s]ign will always lead to sign, one substituting the other...as signifier and signified in turn" (1976, p. xix). In Derrida's own words, knowledge must be understood as a field "'of *freeplay*, that is to say, a field of infinite substitutions'" (Spivak, 1976, p. xix) in which identity fragments into sheer difference, *différance*. The specific way this takes place is by a recursive process of self-referential mirroring in which, time and again, the signifier/subject is displaced by being made into the signified object of a newly implicit subject. Therefore, if we view the self-reflection of modernism as a mirroring that maintains in abstraction the classical relation of object-in-space-before-subject, postmodernism would then constitute an infinite repetition of this mirroring, one that maintains classical identity in such a way that, in the end, it also negates it, exposing the "black hole" at its center. In the "field of infinite substitutions" constitutive of the poststructuralist text, in the infinite regress of signs within signs within signs, lies the addict's inability to let go of the deeply ingrained habit of ontical comportment. The sterility of the "freeplay" is certainly not lost on the postmodernists. In their abstract self-reflections, there is the distinct mood of disenchantment and deflation that constitutes the "down phase" of the modernist/postmodern roller coaster ride. Yet, because the postmodern writer can find meaning nowhere else, the rule of reflection lingers on, albeit in this negative, thoroughly self-subverting manner.

The psychologist and philosopher Eugene T. Gendlin articulates a phenomenological alternative to both modernist and postmodern writing in his expansion upon Heidegger. Gendlin (1993) offers us a text that can reflect upon its bodily, prereflective source. "Speaking is a special case of...bodily living," says Gendlin. "[O]ur bodies perform the implicit functions essential to language....Our bodies imply our...linguistic meanings" (p. 34). Moreover, when we speak or write, the prereflective source of this activity is not simply left behind; it continues to operate in the very midst of our linguistic functioning. Thus, for example, "the most sophisticated details of a linguistic situation can make our *bodies* uncomfortable" (1993, p. 34). Could we not reflect explicitly upon the prereflective source of our reflection? Let us attempt such an act of self-reflection here, with the very words on this page. If Gendlin is correct, our reading of these words arises from our bodies, and since this bodily source goes on functioning even as our words now turn back upon it, it seems we should be able to realize that source in a bodily way so that our words no longer appear as mere abstractions. This is what Gendlin means when he proclaims that "words can say how they work" (1993, p. 29): they work from the body, and, becoming cognizant of their own bodily underpinning, they can link back to it. As in Gendlin's approach, the words of the postmodern text do reflect upon themselves, but only as disembodied signs ultimately devoid of meaning. Gendlin points us beyond postmodernism. In Gendlin's form of self-reflection, the text is not merely conscious of itself *as* an abstract text, but calls attention to the concrete process from which it originates. Only by gaining access to this prereflective "subtext" can we supersede the old ontical trichotomy of object-in-space-before-subject and concretely approach the (w)holeness of Being. In Gendlin's terms, the prereflective is "pre-separated" (1991b, pp. 116–117); that is, it "comes before," is more primordial than the divisions arising in classical thinking and perpetuated in modernism and postmodernity.

We must be clear about the order of embodiment we need to access if *Being* is to be recognized as our "subtext." The body we must Proprioceive is no finite particular flesh but the *flesh of the world.* To repeat Merleau-Ponty's portrayal of this,

> The flesh is not matter, in the sense of corpuscles of being...is not mind, is not substance. To designate it, we should need the old term "element," in the sense it was used to speak of water, air, earth, and fire, that is, in the sense of a *general thing*, midway between the spatio-temporal individual and the idea, a sort of incarnate principle that brings a style of being wherever there is a fragment of being. The flesh is in this sense an "element" of Being. (1968, p. 139)

Defining the flesh in this way, Merleau-Ponty invokes the ontological difference. The flesh of the world is not merely a "spatio-temporal individual," not merely a *b*eing; it is "an 'element' of *B*eing." What of the flesh of *this text*? To apprehend it, we cannot limit ourselves to the realm of the senses but must work in the medium of language and thought. For Merleau-Ponty, the cognitive dimension functions as a "*second* flesh": "It is as though the visibility that animates the sensible world were to emigrate, not outside of every body, but into another less heavy, more transparent body, as though it were to change flesh, abandoning the flesh of the [sensible] body for that of language" (1968, p. 153). It is this *bodily* character of language and thinking that is denied when thinking is geared to move irreversibly "forward," from the Cartesian subject, to its object. And the Proprioceptive reversal of thinking would amount to a withdrawal of the projections of the Cartesian *cogito*, a "*retro*jection" that would disclose the actual grounding of Cartesian consciousness in the "second flesh"—in the bodily Being of word and thought. Once again, the true author of this text is three-dimensional Being; it is s/he that governs the writing in which I now am engaged. Therefore, it is s/he who must be Proprioceived. Short of doing this, no matter what I say *about* Being, the detached *cogito* will continue to preside.

Evidently then, to foster the requisite Proprioception, I must reverse gears and move backward into the generic body that is composing these words I am typing. But something is blocking my way. Though I wish to engage the *generic* authorial body, I am confronted with the compelling sense that the body behind these words is only that of *Steven Rosen*. The way he is normally experienced (by himself and others), Steven is clearly just an individual; an ontically constituted entity; a particular being who is defined by his personal needs and personal history—dating back to the constitution of his body image around the time of the mirror stage. Steven's particular subjectivity takes form as an *objectification* of "second flesh" that obscures the generic thinking body and thus promotes the rule of the abstract *cogito*.

Let me try to put it in more experiential terms. For these words that I type to "say how they work," it seems I must somehow enter my body and gain a concrete sense of the manner in which the words are produced. To begin, I might simply bring my attention to my hands and fingers as they move across the keyboard, obtaining a kinesthetic feel for the working of my muscles. Such proprioceptive awareness can readily be achieved. But there is one thing "I" normally do *not* proprioceive, namely, *this "I" itself.* Merleau-Ponty called the "I" "that central vision that joins the scattered visions, that unique touch that governs the whole tactile life of my body as a unit, that *I think* that must be able to accompany all our experiences" (1968, p. 145). This "I" is the prime "identity operator" (to adapt a term from mathematics) upon which all of "my" specific operations are based; it is the "master word" that lies behind all these particular words. The "I think" is what "I" must Proprioceive to gain access to the generic thinking body; that is, Being must think itself.

Approaching the "I" in the forward direction will not suffice. It is when the "I" is *con*-ceived rather than Proprioceived that generic Being becomes the finite particular being of "Steven Rosen." What is the effect of unquestioningly presupposing "Steven's" personal identity in seeking to engage the body behind these words? A certain kind of closed container is tacitly created for the bodily process "I" become aware of, since it is taken for granted that said linguistic process is confined within a particular body, that of the "spatio-temporal individual," "Steven Rosen." When generic Being is thus narrowed into "Steven Rosen's" personal being, Being is occluded. Entering the body in this ontical way, by no means is the *cogito's* own body felt from within. On the contrary, the *cogito* remains anonymous; it maintains the disembodied from-to stance it has held at least since the time of Descartes. So "I," "Steven Rosen," may feel "my" way backward into "my" body to gain a sense of what is happening within "me" as "I" type these words, but the "inner flesh" thereby engaged will not be the flesh of the world. To encounter generic flesh, ontical proprioception is not enough; only a Proprioception of

Being as such will accomplish the task. We might say that Rosen's particular flesh *is* the flesh of the world, but, in the objectification carried out by recognizing only the former order of embodiment, the latter will continue in obscurity.

Just what sort of practice would be required for the Proprioception that is needed? In what specific way would our apprehension of this text have to change in order that we consciously engage three-dimensional Being per se? Yes, it is generally true that, to reach the apeironic ground of these words about Being, the "master word"—the "I think"—must itself be Proprioceived. David Bohm's notion of "proprioceptive thought" is relevant in this regard. What I would now add to what I said about this in the previous chapter is that such a reversal of thinking's gear would have to be grounded in the generic *body*. But can anything more be said about just how we are to enter this body? In Gendlin's experiential approach, we are advised to enter the body through the middle: "Let your attention refer inside, directly, physically, to the comfort or discomfort in the middle of your body" (1991a, p. 45), say, in your "chest [or] stomach" (1996, p. 1). The alternative I suggest so as to facilitate the Proprioception of the *thinking* body that governs this ontological writing is that we enter the body through the *head.* My proposal is based on the work of the American social psychiatrist, Trigant Burrow.

In the foregoing chapter I noted that, years before Bohm, Burrow had spoken of the need for human beings to gain proprioceptive awareness of the organismic basis of thinking and language, these being the activities Burrow deemed responsible for the fragmentation of human society. Burrow's term for the "I" anonymously engaged in ceaseless objectification is the *'I'-persona.* Importantly, the functioning of the 'I'-persona has a distinct anatomical locus. It is centered in what Burrow called the "cerebro-ocular" region (1953, p. 526), that is, in the cerebral cortex of the brain and in the organ of vision associated with it. Burrow pointed out that it was through the phylogenetic development of the cerebral cortex that language and symbolic activity first arose. Therefore, to gain an immediate sense of this activity, it seems one would have to "enter the body" through the cerebrum. But this conclusion was informed by more than a simple logical deduction. Burrow claimed to have had a spontaneous experience of the 'I'-persona's bodily base, one that profoundly influenced all his subsequent research. After a prolonged period of inter-personal strife involving the members of the group that he had established to investigate such "I"-based conflict, he began to notice a distinctive pattern of tension around his eyes and forehead. Burrow recognized in this the concrete expression of the 'I'-persona.

I have proposed that, to enter the climactic stage of Ontogeny, we must engage in Proprioception, which means consciously receiving the gift of thinking that organic Being has given to itself. In Burrow's terms, this entails bringing attention—not so much to sensations in the chest or stomach, or in the hands—but to the ocular-facial or "cephalic segment" (1953, pp. 249–254), that is, to the area of the body around the forehead and eyes. Burrow would caution us not to confuse the 'I'-persona that resides therein with the ego of the allegedly isolated individual. We might say that this persona is the *species-wide* "subject" that lies behind the appearance of individual subjectivity—the subjectivity of "Steven Rosen," for example. But while it is through the 'I'-persona that we, as a species, create the impression of ourselves as merely isolated, disembodied subjects, the generic "I" itself is no disembodied subject. Instead, it is the *bodily process* that is central to human functioning as a whole. Therefore, when Burrow became attentive to the 'I'-persona rather than continuing to be unwittingly governed by it, he experienced this palpable pattern of tension around the eyes and forehead against the "tensional pattern of the organism as a whole" (Galt, 1995, p. 31). He was thus presumably able to apprehend in an immediate way what he called the "solidarity of the species" (Burrow, 1953, p. 71)—what Merleau-Ponty called the "flesh of the world"—what we have called *apeironic Being.*

Following his first spontaneous Proprioception of the generic organism, Burrow sought to cultivate the experience in a systematic practice he named "cotention" (Burrow, 1932). He described his procedure as one of setting aside daily experimental periods in which he "adhered consistently to relaxing the eyes and to getting the kinesthetic 'feel' of the tensions in and about the eyes and in the cephalic area generally" (1953, p. 95). Burrow might say that—behind "Steven Rosen's" work with this text, behind the typing of these very words—the master word or 'I'-persona operates, and that this generic "I" manifests itself concretely in the muscular activity of "Steven's" eyes. It may well be easier to proprioceive movement in other parts of the body than in the eyes; since ocular activity is subtler, a great deal more practice may be necessary. Nevertheless, if Burrow is right, it is proprioception of the activity in and around "my" eyes that would be required for generic Proprioception. So it seems that, if "I" am to go beyond functioning exclusively as an "abstract head," "my" practice evidently would have to include obtaining a bodily sense of the very "head" that now directs this writing.

As "I" write these words then, I am to turn my awareness to the tensions in my eyes behind which lies the "master word," the "I." However, such a practice apparently would still not suffice. For it seems that, if I were to limit myself to this, a decisive gap would yet remain between the abstract meanings of these words and their corporeal base. What more would be needed to close this hiatus between my reflection upon the prereflective and prereflective Being itself? I propose that the Proprioceptive circle can be closed, that the spirit of these words can wholly be merged with their material ground, if the sign for the master word (in English, the letter "I") is *topologically transformed.*

First, let me emphasize once more that this post-postmodern text must be paradoxical in character. Again, what we are seeking to do is include in this reflection of ours the prereflective source of our activity, the ontological "subtext" that normally is kept strictly implicit. Our text—whose signifiers stand in for ourselves (for "me," who writes these words, and for "you," who reads them)—is to draw back in upon itself like Cézanne's "auto-figurative" painting, make reference to itself without alienating itself, as happens with modernist and postmodern texts; in so signifying itself, this text cannot merely turn itself into an other that is cast before a newly implied, more abstract self. Does this mean that the self that is signified must be the *same* self that is doing the signifying? Not exactly. If the self in question were simply the same, our reflection would collapse into mere self-identity. As long as we are engaged in reflection, are working with a text, are writing or speaking and not just remaining silent, there can be no simple self-identity. Yet, even though the very act of reflecting upon the self turns it into what is other, it is possible for this other to flow right back into the source from which it arises, rather than appearing *merely* as an other cast before a new self. Thus, the self-reflection I am describing would give us neither self nor other, in the categorial sense of these terms. We would realize instead their paradoxical interpenetration. And this apeironic paradox is what we require to supersede the supremacy of reflective predication. Signifier and signified would be more than reciprocally interdependent in such a self-reflective text. They would be identical, utterly one. Yet they also would be two. By virtue of the latter aspect, reflection would continue; by virtue of the former, the "pre-separated," prereflective dimension would be brought into play.

However, for the gap between reflection and the prereflective thus to be closed, the paradoxical return of signification to itself requires a signifier that possesses *sufficient dimensionality*. Enter here the Klein bottle. Our work with this structure clearly is no exercise in "pure mathematics" in which a mathematical object is signified by a definition or algebraic formula. For us, the Klein bottle is not merely a signified object; it is a signifier that paradoxically signifies itself. This is the mode of signification that Derrida refused, and that Heidegger and Merleau-Ponty indicated only vaguely (pre-topologically). It is the *dimensionality* of the Kleinian signifier that permits us to close the gap between the reflective and prereflective.

We have found in general that classical reflection takes place in a context that is assumed to be three-dimensional but that, in fact, is fractionally less than three. And I have proposed that full three-dimensionality can be realized dialectically, not by a higher order extrapolation of reflective dimensionality, but by engaging the prereflective dimension of Being. Of course, this bare statement about the prereflective in itself is merely reflective; the "prereflective dimension" is the content of our predication but that abstract content is simply contained as what we reflect upon in the closed linguistic space of our discourse; from here, we cannot pass smoothly to concrete prereflectivity itself. This limitation is related to the fact that the signifiers we use to convey our three-dimensional content themselves lack sufficient dimensionality. By themselves, these words, these strings of one-dimensional marks printed on this two-dimensional surface, are indeed inadequate to make a reality of apeironic paradox; they are but arbitrarily devised, conventionally agreed upon tokens that refer to their content in a merely external manner. The dimension of these signifiers is to be raised, but a preliminary step is of course required. We must shift our attention to the *master* word, work with the one-dimensional "I."

Signifying the "I" paradoxically, we say, "*I am not-I.*" This is no mere affirmation or denial of a predicated content, but predication's denial of itself. In asserting that "I am not-I," the customary subject/predicate format *is* being used ("I" is the subject, "am not-I" is the predicate), but in a manner in which the content this sentence expresses calls the form into question. The paradoxical statement amounts to a declaration that the syntactical boundary condition that would delimit "I" cannot effectively do so; simple predicative boundary assignment is thwarted, so that even though "I" implicitly is being posited as distinct from that which is external to it, at the same time, it is inseparable from it. The statement "I am not-I" tends to boggle our minds because the human mind, when operating ontically, is solely a discriminatory organ. Indeed, it is that prereflectively chosen "I" whose principal function is to reflect—to posit, draw boundaries, finitize, separate. In seeking the *ontological*, "I am not-I" is the paradoxical self-signification that is to be fleshed out by raising the dimension of the signifier.

We began Chapter Four with Heidegger's designation of Being as "the presencing of what is present" (1946/1984, p. 39). This is paradoxical, since "Being…ambiguously names what is presently present and also…what is absent" (p. 37). Thus Being is, and it is not. Or, in terms of our "master word," *I am not-I.* In the concretization of Being carried out in previous sections of the present chapter, we went from mere words of Being to its Necker cube, lemniscate, and Moebius strip expressions. With these steps, the dimensionality of the signifier was already being increased and our expression of ontological paradox was becoming more embodied. But, up to that point, we were still working with mere symbols, reflectors whose prereflective content in fact remained abstract, being contained within our field of reflection. Only with the fully three-dimensional *Kleinian* reflector can we surpass the primacy of reflection, for only the Klein bottle signifies Being by signifying *itself*. Therefore, instead of just saying "Being" or "I am not-I," we can let the Klein bottle do our speaking. It can become our text. It should be clear that what would make this text "Kleinian" is not merely its signification of the Klein bottle in words, schematic drawings, or objective models in space. By "Kleinian text," I mean a text in which the Klein bottle serves as the medium for dimensionally extending these self-reflective words (in particular, "I am not-I") so that they can actually reenter their own prereflective ground. Here the higher-dimensional Klein bottle is embodied, not just as an object of signification, but as part of the signifying act itself.

It is the uroboric movement of the Klein bottle back into itself that produces the hole in it. Since this opening is the mark of self-signification, with it, the classical gap between signified and signifier in fact should be closed. With the hole in the Kleinian text, we should at once have wholeness. However, this will not happen if we continue to assume a reflective posture in our relation to the text; in that case, the (w)hole will appear as *merely a hole*, a breach in the text that renders it incomplete. Clearly then, we must change our posture, approach the Kleinian text in a different way. Specifically what way? How can we overcome the old, compelling tendency to turn whatever we signify into an object of reflection appearing "out there" before us, simply contained within its context and thus set apart from us? What must we do to allow the self-containing Kleinian signifier to signify itself by flowing right back into us? We already know the answer.

I am proposing that the Kleinian text in truth cannot be simply contained. In containing itself, the Klein bottle should spill over the bounds of the context that would enclose it, flow backward to its own prereflective ground—*our* ground, we who read this text. Yes, as Gendlin would say, the bodily source of these words continues to operate as we read them. And, with the dimensional enhancement of these paradoxical words of Being that fleshes them out, makes them concrete, the gap between this reflection on the prereflective and its living source should indeed be closed. Thus, in properly completing our Kleinian discourse, this text we read would live within us as it stands before us, and would do so without interruption. And we already have an indication of how to work with the text so as to complete it in this way.

Although the long-prevalent habit of classical reflection strongly disposes us to approaching the Klein bottle in the "forward gear," that is to say, as but an object of predication, this "object" does not lend itself to being predicated thus. The inherent character of the Klein bottle suggests that we "reverse gears." Instead of approaching the ontological text in an exclusively ontical manner via *conceptual* activity, we are to employ *Proprio*ception. The way Gendlin would put it, we are to obtain a "moody understanding" (1993, p. 30) or "felt sense" (1978) of our Kleinian text, a bodily cognizance that exceeds the text as a mere content we reflect upon; the "felt sense," of course, is the Proprioceptive awareness of the prereflective subtext. It should be true that we could gain such a sense of *any* text, since all texts originate in the prereflective. But when the text appears closed into its context and thus closed off from its prereflective source, the gap between the reflective and prereflective will persist. In the case of the Kleinian text, there can be no pretense that it is categorically divided from its subtext, for, *as* a text, it is incomplete. To complete it, we must follow its own natural trajectory backward into the living matrix that constitutes the origin of our reflection upon it. In this way, the Kleinian text comes alive, stands within us as well as before us. Our task is to surpass the long dominant habit of predication so that—in one unbroken movement—we may pass from a reflective understanding of the prereflective ground of this text to the ground itself. The Klein bottle uniquely mediates this backward passage. The hole in this Kleinian text is the "perfect size and shape" for our inwardness, and, like a black hole in space, it draws us toward it, extending to us a pregnant invitation to fill it with the whole of "our selves"—that is, to fill it with a Proprioceptive awareness of *apeiron*. This is how the Black Goddess brings light to herself.

The Kleinian text thus responds to the call to Being by acknowledging Being and fleshing it out. Instead of taking Being as an unacknowledged point of departure and moving forward from there into beings, a backward movement into Being per se takes place. In Heidegger's terms (1954/1968), a new kind of thinking occurs, one that thinks the very source of thinking itself—thinks Being—and, in so doing, gratefully receives what Being has given in self-Appropriation. While the Klein bottle lends itself to such self-thinking in a way that no ordinary object could, the thinker must still be prepared for the "reversal of gears" that makes this possible. If not, the necessary hole in the Kleinian text will be misconstrued, taken as *merely* a hole. I am proposing then that Proprioception is the appropriate way to work with the Klein bottle, and that it is what the self-containing Klein bottle requires and invites. In thinking this Kleinian text, we must think Proprioceptively, think our own thinking.

We know exactly what this means in terms of concrete practice. Following Burrow (1932), we must practice *cotention*. The bodily locus of the "heady" linguistic activity we are involved in indeed is the *head*. Therefore, in proceeding Proprioceptively, we must gain a bodily sense of our heads. More precisely, in moving backward into the body that is engaged with this Kleinian text, we are to apprehend the dimensionally enhanced "I"—the generic word that governs all these particular words—by obtaining a felt sense of our *eyes*. To echo the words of Merleau-Ponty, the "I" is "that central vision that joins the scattered visions,...that *I think* that must be able to accompany all our experiences" (1968, p. 145). The implication of the close relationship that Burrow established between the visual system and the cognitive activity of the cerebrum is that, by bringing attention to our eyes, the "I" or *cogito* that presides over the working of this Kleinian text can be grasped Proprioceptively. In the process, the hitherto detached and disembodied *cogito* is transmuted into ontological flesh.

Recall Leder's (1990) observation that, ordinarily, in whatever we see, we "cannot see our own seeing" (p. 17), since our seeing is what we see *from*. In the customary "from-to" posture, the "from" is a "nullpoint" (p. 13), a mere point of departure for experience, one that itself is *not* experienced. The cotentive practice entails *reversing* the forward orientation. We are to *arrive* at our "point of departure," *enter into* the previously inconspicuous nullpoint, hole, or "blind spot" in the field of vision. And this "retraction of the light cones" into the point-origin of the visual field, this backward movement of the eyes, brings us to the "central vision" or "I" that functions subtextually to orchestrate this whole text. Proceeding in this manner, "we" are poised to think Proprioceptively; to think "our" own thinking. That is, Being is poised to realize its authorship of the Kleinian text.

Note that Bohm (1994) regarded "proprioceptive thought" as a kind of meditation. For us such "meditation" would not entail a realization of the self that cleanly transcends all symbolic operations; it would not lead to a state of "pure consciousness" in which all relations to objects are left behind. The Proprioception I am speaking of is not undertaken *instead* of perception and conception; symbolic reflection continues; it is this that we Proprioceive. And while our Proprioception of the Kleinian reflector never brings sheer transcendence of it, we do fluidly encompass its prereflective origin. The necessary hole in our Kleinian text calls neither for a continuation of symbolic mediation by itself, nor for the meditative immediacy of "pure experience"; rather, it invites what we may call *medi(t)ation*: the mutual permeation of externally mediated activity and internal self-realization, of reflectivity and the prereflective. Thus cotentive Proprioception would not purely be a meditation upon the self, but a thinking of other that flows right through the hole in this Kleinian other and back into itself. Only in such a process, where we would not simply be seeking to break away from symbolic reflection, could we truly surmount its pervasive influence. And while the reflective would no longer reign supreme, through Kleinian Proprioception it actually would reach culmination, for no longer would it conceal the prereflective ground that sustains it so as to maintain the false impression of being complete unto itself. In recognizing its own incompleteness as a simply autonomous entity, in acknowledging its roots in the prereflective lifeworld, symbolic reflection would gain *authentic* completion. The Proprioceived Kleinian reflector would bring reflection to fruition by bringing to light the aspect of it that previously had not been reflected upon. Put in alchemical terms, the inside-out Kleinian vessel would be *bene clausum* (hermetically sealed): not only entirely open to its own prereflective source but—*because* of this openness—also entirely closed, complete unto itself, fully three-dimensional.

3. Conclusion

In making the (w)holeness of Being a concrete reality, we are to read this text Proprioceptively; read our own reading; read these words about passing beyond themselves (into their prereflective roots) in such a way that the passage actually takes place. Such a reading is mediated by fleshing out this text via the dimensional amplification provided by the Klein bottle. Now the body of our text does not consist merely of lifeless, empty symbols, intrinsically meaningless signifiers that can only point outside themselves to disembodied meaning; our text is the Klein bottle. It is in cotentively reading the (w)hole in this self-containing three-dimensional text (the "blind spot" in the field of vision) that we should pass unbrokenly into its subtext. The Klein bottle's incompleteness when read ontically is at once an incompleteness in our reflecting upon it. In accepting the invitation to Proprioceive the Kleinian text, we complete our reflective symbolic activity by circling back into its prereflective origin.

But old habits do persist. *Can* we read our own reading? Can we read the hole in the Klein bottle in such a way that we relax the compulsion to regard it as *merely* a hole, a gap in an ordinary object simply contained in space? Can we read the hole in the Klein bottle as an opening to "another dimension" and read that dimension as the prereflective source of our very own reading? Can we *enter* that dimension through Proprioception? It is a matter of proceeding medi(t)atively, from "both sides at once"—from the mediative side of the conceptual, and from the side of experiential immediacy known in meditation. The Kleinian concept brings us to the limit of the conceptual. Because the true boundary of our symbolic activity must be paradoxical, a boundary that is *not* a boundary (lest we continue in the boundary-making mode of symbolic reflection), it is precisely at this inner horizon that the conceptual and the experiential can flow into one another without interruption. It is here that we can realize the intimate harmony of outside and inside, of object and subject, of the bounded and boundless. That is to say: the harmony of *apeiron* can be realized in full by the Goddess herself.

EPILOGUE

In this book I have sought to explore at its roots the fragmentation of human culture so pervasive at this juncture of our history. I began the Preface by reading the litany of woes currently besetting humankind. These problems are all too well known, and I will not repeat my recitation of them here. In summing up, what I want to reiterate is what I believe to be the crux of our dilemma, and the sea change in human functioning that is needed to address it.

Why is it that, despite the good intentions of a great many highly intelligent, talented, and inventive individuals throughout the world, our problems only seem to be increasing, and in fact appear to be rapidly reaching ominous proportions? I propose it is essentially because, however well-meaning we have been, what "we" have never fully questioned is this very *we*. That is, in our valiant efforts to remedy our grave situation by achieving an insight into it, we have failed to gain sufficient insight into the prime players in the drama: ourselves. But haven't the more reflective members of our species been searching their souls from time immemorial? It is indeed true that, ever since the rise of reflective consciousness, we have been seeking to know ourselves—as Socrates exhorted over two thousand years ago. This is what our quest for individuation has been all about. What I am suggesting, however, is that this quest has always been frustrated because human reflectivity has always been geared to turn whatever it seeks to know into an *object*, including itself. In this way, it has alienated itself, even as it has striven for self-intimacy. It is our self-alienation then, I submit, that has brought us to the brink of catastrophe.

In the present book, I have proposed that the self-knowledge required for grasping the true nature of our predicament demands that we take an unprecedented step. We must effect a radical "reversal of gears." The reflective individual must break the centuries-old habit of moving away from himself toward his object, must move backward into himself to the prereflective source of his reflection. There he will find that he is not merely a free-standing subject after all, nor is he merely an object. Instead, "he" is the *embodied fusion of subject and object that constitutes the paradox of apeiron.* So—if effectively addressing humankind's current crisis means gaining self-knowledge—it is *apeiron* we must come to know.

In the present volume, knowledge of apeironic Being is by no means complete. I intimated in Chapter Four that the "motherly" self-birthing of three-dimensional Being gains "midwifely" support from a "fourth dimensional" source referred to by Heidegger as *Appropriation* (*Ereignis*). Having employed topology in Chapter Five to flesh out the action of three-dimensional Being, might we not be able to flesh out the relationship between this apeironic mother and her other-dimensional midwife through some further elaboration on topology? Although detailed consideration of this question is beyond the scope of the present volume, I have investigated the matter. It turns out that there are *several* dimensions of apeironic Being, that these correspond to several interwoven topological structures, and that the different dimensions of *apeiron* evolve together in an intricate pattern of individuation that far surpasses the Ontogeny of three-dimensional Being alone. Moreover, these co-evolving orders of topo-dimensional flesh constitute an entire ecosphere, so to speak. For, while the three-dimensional "flesh…of language" (Merleau-Ponty, 1968, p. 153) embodied by the Klein bottle is limited to the human sphere, the other dimensions bring into play the lifeworlds of animal, vegetable, and mineral. An account of these additional dimensions requires nothing less than a volume unto itself to achieve coherent expression (see Rosen, 2004). Suffice it to say for now that the self-realization of three-dimensional Kleinian Being described in the previous chapter is certainly not the final word on dimensional development. The Kleinian structure opens not only inwardly into its own ground, but beyond itself to other-dimensional embodiments of *apeiron*.

Let me underscore the point metaphorically by concluding this book on an alchemical note. Near the end of Part One, I observed that the fashioning of alchemy's Hermetic vessel symbolizes the process of individuation that culminates in the *unio mentalis* (Jung, 1970, p. 465), a state of intellectual maturity that brings mental development to fruition. But Jung noted that "the alchemists rightly regarded 'mental union in the overcoming of the body' as only the first stage of conjunction or individuation....In general, the alchemists strove for a *total* union of opposites" (1970, pp. 474–475). This meant that, once mental integration was attained, there would need to be additional "distillations," alchemical processes entailing a reunion of the now concretely realized mind with the more embodied orders of nature that had been repressed. Given the relationship between the hermetic vessel and the Klein bottle indicated in the last chapter, we might associate the Proprioceptive completion of this topological structure with the *unio mentalis*. To go further, to achieve "a *total* union of opposites," Proprioception evidently must go beyond the mental sphere of Kleinian self-Appropriation to the fleshier orders of topological Appropriation. In this way, the hitherto concealed, nonhuman dimensions of Being would be brought to light and *apeiron* would come to know herself more fully in her multi-dimensional diversity. It is through this ecological self-knowledge that the fragmentation of our whole planet would be surpassed and its (w)holeness reach fruition.

WORKS CITED

Abram, David. (1996) *The Spell of the Sensuous*. New York: Vintage.

American College Dictionary. (1968) New York: Random House.

Angeles, Peter A. (1981) *Dictionary of Philosophy*. New York: Barnes & Noble.

Barr, Stephen. (1964) *Experiments in Topology*. New York: Dover.

Barrett, William. (1958) *Irrational Man*. New York: Doubleday/Anchor.

Berger, John and Jean Mohr. (1982) *Another Way of Telling*. New York: Pantheon Books.

Berman, Morris. (1989) *Coming to Our Senses*. New York: Bantam Books.

Bohm, David. (1951) *Quantum Theory*. Englewood Cliffs, N.J.: Prentice-Hall.

———. (1952) "A Suggested Interpretation of the Quantum Theory in Terms of 'Hidden' Variables, I and II," *Physical Review*, 85:2, pp. 166–193.

———. (1980) *Wholeness and the Implicate Order*. London: Routledge and Kegan Paul.

———. (1986a) "A New Theory of the Relationship of Mind and Matter," *American Journal of Psychical Research*, 80: 2, pp. 113–136.

———. (1986b) "Time, the Implicate Order, and Pre-Space." In *Physics and the Ultimate Significance of Time*, ed. David Ray Griffin. Albany, N.Y.: State University of New York Press, pp. 177–208.

———. (1994) "The Bohm/Rosen Correspondence." In *Science, Paradox, and the Moebius Principle*, ed. Steven M. Rosen. Albany, N.Y.: State University of New York Press, pp. 223–258.

Burrow, Trigant. (1932) *The Structure of Insanity*. London: Kegan Paul, Trench, Trubner, and Co.

———. (1953) *Science and Man's Behavior*. New York: Philosophical Library.

Čapek, Milič. (1961) *Philosophical Impact of Contemporary Physics*. New York: Van Nostrand.

Caputo, John. (1997) *The Prayers and Tears of Jacques Derrida: Religion Without Religion*. Bloomington, Ind.: Indiana University Press.

Carter, Brandon. (1968) "Global Structure of the Kerr Family of Gravitational Fields," *Physical Review*, 174, pp. 1559–1571.

Cataldi, Susan L. (1993) *Emotion, Depth, and Flesh: A Study of Sensitive Space*. Albany, N.Y.: State University of New York Press.

Cixous, Hélène. (1980) "The Laugh of Medusa." In *New French Feminisms*, eds. Elaine Marks and Isabelle de Courtivron. Brighton, Sussex: Harvester Press, pp. 245–264.

Cornford, Francis M. (1957) *Plato's Cosmology*. New York: Liberal Arts Press.

Crease, Robert P., ed. (1993) *Proceedings, Heidegger Conference*. Stony Brook, N.Y.: Department of Philosophy, State University of New York at Stony Brook.

Cunningham, Lawrence S. and John J. Reich. (1990) *Culture and Values*. New York: Holt, Rinehart, and Winston.

Derrida, Jacques. (1967/1982) *Margins of Philosophy*, trans. Alan Bass. Brighton, Sussex: Harvester Press.

———. (1976) *Of Grammatology*, trans. Gayatri C. Spivak. Baltimore: The Johns Hopkins University Press.

———. (1995) *On the Name*, trans. David Wood, John P. Leavey, and Ian McLeod. Stanford, Cal.: Stanford University Press.

Eddington, Arthur Stanley. (1946) *Fundamental Theory*. Cambridge: Cambridge University Press.

Edie, James M., ed. (1964) *The Primacy of Perception*. Evanston, Ill.: Northwestern University Press.

Eisler, Riane. (1987) *The Chalice and the Blade*. San Francisco: Harper & Row.

Ernst, Bruno. (1986) *Der Zauberspiegel des M.C. Escher*. Berlin: Taco.

Everdell, William R. (1997) *The First Moderns*. Chicago: University of Chicago Press.

Fabricius, Johannes. (1976) *Alchemy*. London: The Aquarian Press.

Felser, Joseph. (1999) "From Darkness to Light: A Philosophical Musing on the Hanukkah Myth, the Return of the Goddess, and the End of Religion," *Mythosphere*, 1:4, pp. 463–507.

Foucault, Michel. (1972) *The Archeology of Knowledge*, trans. A.M. Sheridan Smith. New York: Pantheon Books.

———. (1975) *The Birth of the Clinic*, trans. A.M. Sheridan Smith. New York: Vintage Books.

Fox-Keller, Evelyn. (1985) *Reflections on Gender and Science*. New Haven, Conn.: Yale University Press.

Freud, Sigmund. (1914/1957) "On Narcissism: An Introduction." In *Standard Edition of the Complete Works of Sigmund Freud*, Volume 14, ed. James Strachey. London: Hogarth Press, pp. 117–140.

Fuller, B.A.G. and Sterling M. McMurrin. (1957) *A History of Philosophy*. New York: Henry Holt and Co.

Galbraith, Mary and William J. Rapaport, eds. (1991) *Subjectivity and the Debate over Computational Cognitive Science*. Buffalo, N.Y.: Center for Cognitive Science, State University of New York at Buffalo.

Galt, Alfreda. (1995) "Trigant Burrow and the Laboratory of the 'I,'" *The Humanistic Psychologist*, 23 (Spring), pp. 19–39.

Gardner, Martin. (1979) *The Ambidextrous Universe*. New York: Charles Scribner's Sons.

Gebser, Jean. (1985) *The Ever-Present Origin*, trans. Noel Barstad and Algis Mickunas. Athens, Ohio: Ohio University Press.

Gendlin, Eugene T. (1978) *Focusing*. New York: Bantam Books.

———. (1991a) "Crossing and Dipping: Some Terms for Approaching the Interface between Natural Understanding and Logical Formulation." In *Subjectivity and the Debate over Computational Cognitive Science*, eds. Mary Galbraith and William J. Rapaport. Buffalo, N.Y.: Center for Cognitive Science, State University of New York at Buffalo, pp. 37–59.

———. (1991b) "Thinking Beyond Patterns: Body, Language, and Situations." In *The Presence of Feeling in Thought*, eds. Bernard denOuden and Marcia Moen. New York: Peter Lang, pp. 27–189.

———. (1993) "Words Can Say How They Work." In *Proceedings, Heidegger Conference*, ed. Robert P. Crease. Stony Brook, N.Y.: Department of Philosophy, State University of New York at Stony Brook, pp. 29–35.

———. (1996) "Making Concepts from Experience," presented at *International Focusing Conference*, Gloucester, Mass. (2–6 May). Unpublished transcript (33 pp.) available at The Focusing Institute, 34 East Lane, Spring Valley, New York 10977; or http://www.focusing.org.

Gendlin, Eugene T. and Jay L. Lemke. (1983) "A Critique of Relativity and Localization," *International Journal of Mathematical Modelling*, 4, pp. 61–72.

Gombrich, Ernst H. (1960) *Art and Illusion*. New York: Pantheon Books.

Graves, John C. (1971) *The Conceptual Foundations of Contemporary Relativity*. Cambridge, Mass.: MIT Press.

Greene, Brian. (1999) *The Elegant Universe*. New York: W. W. Norton & Co.

Griffin, David Ray, ed. (1986) *Physics and the Ultimate Significance of Time*. Albany, N.Y.: State University of New York Press.

Grosz, Elizabeth. (1994) *Volatile Bodies*. Bloomington, Ind.: Indiana University Press.

Hammer, Anita. (1999) "Mirroring and the Topology of Theatre," presented in Cross-Disciplinary Seminar, Trondheim, Norway (15 February): Department of Social Anthropology, Norwegian University of Science and Technology.

Hardison, O.B. (1989) *Disappearing Through the Skylight.* New York: Viking Press.

Havel, Ivan M. (1998) *Open Eyes and Raised Eyebrows*. Prague: Vesmír.

Heidegger, Martin. (1927/1962) *Being and Time*, trans. John Macquarrie and Edward Robinson. New York: Harper & Row.

———. (1929/1975) "What is Metaphysics?" In *Existentialism from Dostoevsky to Sartre*, ed. Walter Kaufmann. New York: New American Library, pp. 242–264.

———. (1943/1977) "The Word of Nietzsche: 'God is Dead.'" In *The Question Concerning Technology and Other Essays*, trans. William Lovitt. New York: Harper & Row, pp. 53–112.

———. (1946/1984) *Early Greek Thinking*, trans. David F. Krell and Frank A. Capuzzi. New York: Harper & Row.

———. (1954/1968) *What is Called Thinking?*, trans. J. Glenn Gray. New York: Harper & Row.

———. (1954/1977) "The Question Concerning Technology." In *The Question Concerning Technology and Other Essays*, trans. William Lovitt. New York: Harper & Row, pp. 3–35.

———. (1962/1972) "Time and Being." In *On Time and Being*, trans. Joan Stambaugh. New York: Harper & Row, pp. 1–24.

———. (1962/1977) "Modern Science, Metaphysics, and Mathematics." In *Martin Heidegger: Basic Writings*, ed. David F. Krell. New York: Harper & Row, pp. 247–282.

———. (1963/1975) "My Way to Phenomenology." In *Existentialism from Dostoevsky to Sartre*, ed. Walter Kaufmann. New York: New American Library, pp. 234–241.

———. (1964/1977) "The End of Philosophy and the Task of Thinking." In *Martin Heidegger: Basic Writings*, ed. David F. Krell. New York: Harper & Row, pp. 373–392.

Heisenberg, Werner. (1958) *Physics and Philosophy*. New York: Harper & Row.

Holton, Gerald. (1988) *Thematic Origins of Scientific Thought*. Cambridge, Mass.: Harvard University Press.

Husserl, Edmund. (1911/1965) "Philosophy as Rigorous Science." In *Edmund Husserl: Phenomenology and the Crisis of Philosophy*, ed. and trans. Quentin Lauer. New York: Harper & Row, pp. 71–147.

———. (1935/1965) "Philosophy and the Crisis of European Man." In *Edmund Husserl: Phenomenology and the Crisis of Philosophy*, ed. and trans. Quentin Lauer. New York: Harper & Row, pp. 149–192.

———. (1936/1970) *Crisis of the European Sciences and Transcendental Phenomenology*, trans. David Carr. Evanston, Ill.: Northwestern University Press.

Iran-Nejad, Asghar. (1989) "A Nonconnectionist Schema Theory of Understanding Surprise-Ending Stories," *Discourse Processes*, 12, pp. 127–148.

Irigaray, Luce. (1985) *The Speculum of the Other Woman*, trans. Gillian C. Gill. Ithaca, N.Y.: Cornell University Press.

Jahn, Robert and Brenda Dunne. (1984) *On the Quantum Mechanics of Consciousness* (Appendix B). Princeton, N. J.: Princeton University School of Engineering/ Applied Sciences.

Jantsch, Erich. (1980) *The Self-Organizing Universe*. New York: Pergamon Press.

Jaspers, Karl. (1941/1975) "On My Philosophy." In *Existentialism from Dostoevsky to Sartre*, ed. Walter Kaufmann. New York: New American Library, pp. 158–185.

Jones, W. T. (1952) *A History of Western Philosophy*. New York: Harcourt, Brace & World.

Josephson, Brian D. (1987) "Physics and Spirituality: The Next Grand Unification?" *Physics Education*, 22:1, pp. 15–19.

Jung, Carl Gustav. (1956) *Symbols of Transformation: Collected Works*, Volume 5; trans. R.F.C. Hull. Princeton, N.J.: Princeton University Press.

———. (1967) *Alchemical Studies: Collected Works*, Volume 13; trans. R.F.C. Hull. Princeton, N.J.: Princeton University Press.

———. (1968) *Psychology and Alchemy: Collected Works*, Volume 12; trans. R.F.C. Hull. Princeton, N.J.: Princeton University Press.

———. (1969) *The Archetypes and the Collective Unconscious: Collected Works*, Volume 9, Part I; trans. R.F.C. Hull. Princeton, N.J.: Princeton University Press.

———. (1970) *Mysterium Coniunctionis: Collected Works*, Volume 14; trans. R.F.C. Hull. Princeton, N.J.: Princeton University Press.

Kaufmann, Walter, ed. and trans. (1968) *The Portable Nietzsche*. New York: Viking Press.

———. (1975) *Existentialism from Dostoevsky to Sartre*. New York: New American Library.

Kierkegaard, Søren. (1844/1975) "Dread and Freedom." In *Existentialism from Dostoevsky to Sartre*, ed. Walter Kaufmann. New York: New American Library, pp. 101–105.

———. (1846/1975) "Truth is Subjectivity." In *Existentialism from Dostoevsky to Sartre*, ed. Walter Kaufmann. New York: New American Library, pp. 110–120.

———. (1846/1992) "Subjective Truth, Inwardness; Truth is Subjectivity." In *Existentialist Philosophy*, ed. L. Nathan Oaklander. Upper Saddle River, N.J.: Prentice-Hall, pp. 37–43.

———. (1846–1847/1975) "Authority." In *Existentialism from Dostoevsky to Sartre*, ed. Walter Kaufmann. New York: New American Library, pp. 105–110.

———. (1848/1975a) "On His Works." In *Existentialism from Dostoevsky to Sartre*, ed. Walter Kaufmann. New York: New American Library, pp. 87–88.

———. (1848/1975b) "On His 'Mode of Existence.'" In *Existentialism from Dostoevsky to Sartre*, ed. Walter Kaufmann. New York: New American Library, pp. 88–94.

———. (1859/1975) "That Individual." In *Existentialism from Dostoevsky to Sartre*, ed. Walter Kaufmann. New York: New American Library, pp. 94–101.

Krell, David F., ed. (1977) *Martin Heidegger: Basic Writings*. New York: Harper & Row.

Lacan, Jacques. (1953) "Some Reflections on the Ego," *International Journal of Psychoanalysis*, 34, pp. 11–17.

Lauer, Quentin, ed. and trans. (1965) *Edmund Husserl: Phenomenology and the Crisis of Philosophy*. New York: Harper & Row.

Lawlor, Leonard. (1999) "We Need a Name for What We Do: Report on Contemporary Merleau-Ponty Research in the United States," *Chiasmi International*, 1, pp. 27–34.

Leder, Drew. (1990) *The Absent Body*. Chicago: University of Chicago Press.

Levin, David Michael. (1985) *The Body's Recollection of Being*. London: Routledge and Kegan Paul.

Madison, Gary B. (1981) *The Phenomenology of Merleau-Ponty*. Athens, Ohio: Ohio University Press.

Marks, Elaine and Isabelle de Courtivron, eds. (1980) *New French Feminisms*. Brighton, Sussex: Harvester Press.

Matsuno, Koichiro. (1999) "The Clock and its Triadic Relationship," *Semiotica*, 127, pp. 433–452.

Matsuno, Koichiro and Stanley N. Salthe. (1995) "Global Idealism/Local Materialism," *Biology and Philosophy*, 10, pp. 309–337.

Mazis, Glen A. (1987) "'Modern Depths,' Painting, and the Novel," *Soundings: An Interdisciplinary Journal*, 70, pp. 121–144.

McLean, Adam. (1988) "The Alchemical Drama of Goethe's Faust," *Hermetic Journal*, 40, pp. 35–42.

McLuhan, Marshall. (1964) *Understanding Media*. New York: New American Library.

Melucci, Alberto. (1996) *The Playing Self*. Cambridge: Cambridge University Press.

Merleau-Ponty, Maurice. (1962) *Phenomenology of Perception*, trans. Colin Smith. London: Routledge and Kegan Paul.

———. (1964) "Eye and Mind." In *The Primacy of Perception*, ed. James M. Edie. Evanston, Ill.: Northwestern University Press, pp. 159–190.

———. (1968) *The Visible and the Invisible*, trans. Alphonso Lingis. Evanston, Ill.: Northwestern University Press.

Nietzsche, Friedrich. (1872/1967) *The Birth of Tragedy*, trans. Walter Kaufmann. New York: Random House.

———. (1878/1968) "On the Vision and the Riddle." In *The Portable Nietzsche*, ed. and trans. Walter Kaufmann. New York: Viking Press, pp. 267–272.

———. (1880/1992) "Human, All Too Human." In *Existentialist Philosophy*, ed. L. Nathan Oaklander. Upper Saddle River, N.J.: Prentice-Hall, pp. 128–135.

———. (1882/1974) *The Gay Science*, trans. Walter Kaufmann. New York: Random House.

———. (1882/1992) "The Greatest Weight." In *Existentialist Philosophy*, ed. L. Nathan Oaklander. Upper Saddle River, N.J.: Prentice-Hall, pp. 142–143.

———. (1887/1992) "On the Genealogy of Morals." In *Existentialist Philosophy*, ed. L. Nathan Oaklander. Upper Saddle River, N.J.: Prentice-Hall, pp. 117–121.

———. (1889/1968) "Twilight of the Idols." In *The Portable Nietzsche*, ed. and trans. Walter Kaufmann. New York: Viking Press, pp. 463–563.

———. (1895/1968) "The Antichrist." In *The Portable Nietzsche*, ed. and trans. Walter Kaufmann. New York: Viking Press, pp. 565–656.

———. (1895/1975) "The Will to Power." In *Existentialism from Dostoevsky to Sartre*, ed. Walter Kaufmann. New York: New American Library, pp. 130–132.

———. (1895/1992) "The Will to Power." In *Existentialist Philosophy*, ed. L. Nathan Oaklander. Upper Saddle River, N.J.: Prentice-Hall, p. 106.

Oaklander, L. Nathan, ed. (1992) *Existentialist Philosophy*. Upper Saddle River, N.J.: Prentice-Hall.

Ong, Walter. (1977) *Interfaces of the Word*. Ithaca, N.Y.: Cornell University Press.

Ouden, Bernard den and Marcia Moen, eds. (1991) *The Presence of Feeling in Thought*. New York: Peter Lang.

Pioch, Nicolas. (1995) *Picasso and Cubism*. Retrieved 14 September 2000, from The WebMuseum: http://www.tamu.edu/mocl/picasso/wpic/wcubism.html.

Plato. (1965) *Timaeus and Critias*, trans. Desmond Lee. New York: Penguin Books.

Restivo, Sal. (1988) "Modern Science as a Social Problem," *Social Problems*, 35:3, pp. 206–225.

Rosen, Steven M. (1975a) "The Unity of Changelessness and Change," *Main Currents in Modern Thought*, 31:4, pp. 115–120.

———. (1975b) "Synsymmetry," *Scientia*, 110, pp. 539–549.

———. (1980) "Creative Evolution," *Man/Environment Systems*, 10, pp. 239–250.

———. (1983) "The Concept of the Infinite and the Crisis in Modern Physics," *Speculations in Science and Technology*, 6:4, pp. 413–425.

———. (1986) "Time and Higher-Order Wholeness." In *Physics and the Ultimate Significance of Time*, ed. David Ray Griffin. Albany, N. Y.: State University of New York Press, pp. 219–230.

———. (1988) "A Neo-Intuitive Proposal for Kaluza-Klein Unification," *Foundations of Physics*, 18:11, pp. 1093–1139.

———. (1994) *Science, Paradox, and the Moebius Principle.* Albany, N.Y.: State University of New York Press.

———. (1995) "Pouring Old Wine into a New Bottle." In *The Interactive Field in Analysis*, ed. Murray Stein. Wilmette, Ill.: Chiron Publications, pp. 121–141.

———. (1997) "Wholeness as the Body of Paradox," *Journal of Mind and Behavior*, 18:4, pp. 391–423.

———. (1999) "Evolution of Attentional Processes in the Human Organism," *Group Analysis*, 32:2, pp. 243–253.

———. (2004) *Topologies of the Flesh: A Multidimensional Exploration of the Lifeworld.* Unpublished manuscript.

Rucker, Rudolph. (1977) *Geometry, Relativity, and the Fourth Dimension.* New York: Dover Books.

Russell, Bertrand. (1925) *The ABC of Relativity*. New York: Harper & Brothers.

Ryan, Paul. (1993) *Video Mind/Earth Mind: Art, Communications, and Ecology.* New York: Peter Lang.

Sachs, Mendel. (1999) "Fundamental Conflicts in Modern Physics and Cosmology," *Frontier Perspectives*, 8:1, pp. 13–19.

Sartre, Jean-Paul. (1943/1975) "Self-Deception." In *Existentialism from Dostoevsky to Sartre*, ed. Walter Kaufmann. New York: New American Library, pp. 299–328.

Schumacher, John. (1989) *Human Posture*. Albany, N.Y.: State University of New York Press.

Schwartz-Salant, Nathan. (1998) *The Mystery of Human Relationship*. London: Routledge.

Sheets-Johnstone, Maxine. (1990) *The Roots of Thinking*. Philadelphia: Temple University Press.

Sherman, Ernest. (1993) "Merleau-Ponty and the Trickster: Philosophy as the Mytho-Logos of Ambiguity," *Dialectical Anthropology*, 18, pp. 103–120.

Spivak, Gayatri C. (1976) "Translator's Preface." In *Of Grammatology*, Jacques Derrida. Baltimore: The Johns Hopkins University Press, pp. ix–lxxxvii.

Stapp, Henry. (1979) "Whiteheadian Approach to Quantum Theory and the Generalized Bell's Theorem," *Foundations of Physics*, 9:1, pp. 1–25.

———. (1986) "Einstein Time and Process Time." In *Physics and the Ultimate Significance of Time*, ed. David Ray Griffin. Albany, N.Y.: State University of New York Press, pp. 264–270.

Stein, Murray, ed. (1995) *The Interactive Field in Analysis*. Wilmette, Ill.: Chiron Publications.

Strachey, James, ed. (1957) *Standard Edition of the Complete Works of Sigmund Freud.* London: Hogarth Press.

Sundararajan, Louise. (2002) "Shifting Paradigms in the Energy Theory of Emotions: Toward a Synthesis," *Ultimate Reality and Meaning*, 25, pp. 295–306.

Tanabe, Hajime. (1986) *Philosophy as Metanoetics*, trans. Takeuchi Yoshinori. Berkeley, Cal.: University of California Press.

Tarnas, Richard. (1991) *The Passion of the Western Mind.* New York: Ballantine Books.

Vitz, Paul C. and Arnold B. Glimcher. (1984) *Modern Art and Modern Science*. New York: Praeger Books.

Webster's New Collegiate Dictionary. (1975) Springfield, Mass.: G. & C. Merriam.

Webster's New Twentieth Century Dictionary of the English Language (Unabridged). (1976) New York: Collins World.

Werner, Heinz. (1948) *Comparative Psychology of Mental Development*. New York: International Universities Press.

Wittgenstein, Ludwig. (1953) *Philosophical Investigations,* trans. G. E. M. Anscombe. Oxford: Basil Blackwell.

Wolf, Christa. (1984) *Cassandra: A Novel and Four Essays*, trans. Jan van Heurck. London: Virago Press.

Woodman, Marion. (1982) *Addiction to Perfection.* Toronto: Inner City Books.

———. (1996) *Dancing in the Flames*. Boston: Shambhala Publications.

Zukav, Gary. (1979) *The Dancing Wu-Li Masters*. New York: William Morrow.

ABOUT THE AUTHOR

STEVEN M. ROSEN is Emeritus Professor of Psychology at the College of Staten Island of the City University of New York. During his thirty years of teaching and research (1970–2000), he offered courses in both the Psychology and Philosophy Departments. Dr. Rosen currently resides in Vancouver, Canada, where he is actively pursuing interdisciplinary interests that include phenomenological ontology, the philosophy and poetics of science, Jungian thought, the gender question, ecological change, and cultural transformation.

After receiving his Ph.D. in experimental psychology from the City University of New York in 1971, Rosen began to explore the foundations and frontiers of science, his work becoming interdisciplinary and philosophical in nature. He has lectured internationally, and his numerous essays have appeared in a variety of journals and books spanning the fields of philosophy, psychology, education, semiotics, ecology, and theoretical science. Rosen is author of *Science, Paradox, and the Moebius Principle* (Albany, N.Y.: State University of New York Press, 1994) and *The Moebius Seed* (Walpole, N.H.: Stillpoint Publications, 1985).

Dr. Rosen is presently Research Associate and member of the Board of Directors of the Lifwynn Foundation for Social Research, an organization dedicated to carrying forward the work of the American social psychiatrist Trigant Burrow. Rosen is also on the Editorial Board of the *Journal of Mind and Behavior*, and has served as editorial consultant for such journals as *Foundations of Physics* and *Man/Environment Systems*, and for the State University of New York Press.

INDEX

VIBS

The **Value Inquiry Book Series** is co-sponsored by:

Adler School of Professional Psychology
American Indian Philosophy Association
American Maritain Association
American Society for Value Inquiry
Association for Process Philosophy of Education
Canadian Society for Philosophical Practice
Center for Bioethics, University of Turku
Center for Professional and Applied Ethics, University of North Carolina at Charlotte
Central European Pragmatist Forum
Centre for Applied Ethics, Hong Kong Baptist University
Centre for Cultural Research, Aarhus University
Centre for Professional Ethics, University of Central Lancashire
Centre for the Study of Philosophy and Religion, University College of Cape Breton
Centro de Estudos em Filosofia Americana, Brazil
College of Education and Allied Professions, Bowling Green State University
College of Liberal Arts, Rochester Institute of Technology
Concerned Philosophers for Peace
Conference of Philosophical Societies
Department of Moral and Social Philosophy, University of Helsinki
Gannon University
Gilson Society
Haitian Studies Association
Ikeda University
Institute of Philosophy of the High Council of Scientific Research, Spain
International Academy of Philosophy of the Principality of Liechtenstein
International Association of Bioethics
International Center for the Arts, Humanities, and Value Inquiry
International Society for Universal Dialogue
Natural Law Society
Personalist Discussion Group
Philosophical Society of Finland
Philosophy Born of Struggle Association
Philosophy Seminar, University of Mainz
Pragmatism Archive at The Oklahoma State University
R.S. Hartman Institute for Formal and Applied Axiology
Research Institute, Lakeridge Health Corporation
Russian Philosophical Society
Society for Iberian and Latin-American Thought
Society for the Philosophic Study of Genocide and the Holocaust
Unit for Research in Cognitive Neuroscience, Autonomous University of Barcelona
Yves R. Simon Institute

Titles Published

1. Noel Balzer, *The Human Being as a Logical Thinker*

2. Archie J. Bahm, *Axiology: The Science of Values*

3. H. P. P. (Hennie) Lötter, *Justice for an Unjust Society*

4. H. G. Callaway, *Context for Meaning and Analysis: A Critical Study in the Philosophy of Language*

5. Benjamin S. Llamzon, *A Humane Case for Moral Intuition*

6. James R. Watson, *Between Auschwitz and Tradition: Postmodern Reflections on the Task of Thinking.* A volume in **Holocaust and Genocide Studies**

7. Robert S. Hartman, *Freedom to Live: The Robert Hartman Story, edited by Arthur R. Ellis.* A volume in **Hartman Institute Axiology Studies**

8. Archie J. Bahm, *Ethics: The Science of Oughtness*

9. George David Miller, *An Idiosyncratic Ethics; Or, the Lauramachean Ethics*

10. Joseph P. DeMarco, *A Coherence Theory in Ethics*

11. Frank G. Forrest, *Valuemetrics*: *The Science of Personal and Professional Ethics.* A volume in **Hartman Institute Axiology Studies**

12. William Gerber, *The Meaning of Life: Insights of the World's Great Thinkers*

13. Richard T. Hull, Editor, *A Quarter Century of Value Inquiry: Presidential Addresses of the American Society for Value Inquiry*. A volume in **Histories and Addresses of Philosophical Societies**

14. William Gerber, *Nuggets of Wisdom from Great Jewish Thinkers: From Biblical Times to the Present*

15. Sidney Axinn, *The Logic of Hope: Extensions of Kant's View of Religion*

16. Messay Kebede, *Meaning and Development*

17. Amihud Gilead, *The Platonic Odyssey: A Philosophical-Literary Inquiry into the Phaedo*

18. Necip Fikri Alican, *Mill's Principle of Utility: A Defense of John Stuart Mill's Notorious Proof.* A volume in **Universal Justice**

19. Michael H. Mitias, Editor, *Philosophy and Architecture.*

20. Roger T. Simonds, *Rational Individualism: The Perennial Philosophy of Legal Interpretation.* A volume in **Natural Law Studies**

21. William Pencak, *The Conflict of Law and Justice in the Icelandic Sagas*

22. Samuel M. Natale and Brian M. Rothschild, Editors, *Values, Work, Education: The Meanings of Work*

23. N. Georgopoulos and Michael Heim, Editors, *Being Human in the Ultimate: Studies in the Thought of John M. Anderson*

24. Robert Wesson and Patricia A. Williams, Editors, *Evolution and Human Values*

25. Wim J. van der Steen, *Facts, Values, and Methodology: A New Approach to Ethics*

26. Avi Sagi and Daniel Statman, *Religion and Morality*

27. Albert William Levi, *The High Road of Humanity: The Seven Ethical Ages of Western Man*, Edited by Donald Phillip Verene and Molly Black Verene

28. Samuel M. Natale and Brian M. Rothschild, Editors, *Work Values: Education, Organization, and Religious Concerns*

29. Laurence F. Bove and Laura Duhan Kaplan, Editors, *From the Eye of the Storm: Regional Conflicts and the Philosophy of Peace.* A volume in **Philosophy of Peace**

30. Robin Attfield, *Value, Obligation, and Meta-Ethics*

31. William Gerber, *The Deepest Questions You Can Ask About God: As Answered by the World's Great Thinkers*

32. Daniel Statman, *Moral Dilemmas*

33. Rem B. Edwards, Editor, *Formal Axiology and Its Critics.* A volume in **Hartman Institute Axiology Studies**

34. George David Miller and Conrad P. Pritscher, *On Education and Values: In Praise of Pariahs and Nomads*. A volume in **Philosophy of Education**

35. Paul S. Penner, *Altruistic Behavior: An Inquiry into Motivation*

36. Corbin Fowler, *Morality for Moderns*

37. Giambattista Vico, *The Art of Rhetoric* (Institutiones Oratoriae, 1711–1741), from the definitive Latin text and notes, Italian commentary and introduction by Giuliano Crifò. Translated and Edited by Giorgio A. Pinton and Arthur W. Shippee. A volume in **Values in Italian Philosophy**

38. W. H. Werkmeister, *Martin Heidegger on the Way*. Edited by Richard T. Hull. A volume in **Werkmeister Studies**

39. Phillip Stambovsky, *Myth and the Limits of Reason*

40. Samantha Brennan, Tracy Isaacs, and Michael Milde, Editors, *A Question of Values: New Canadian Perspectives in Ethics and Political Philosophy*

41. Peter A. Redpath, *Cartesian Nightmare: An Introduction to Transcendental Sophistry*. A volume in **Studies in the History of Western Philosophy**

42. Clark Butler, *History as the Story of Freedom: Philosophy in Intercultural Context*, with responses by sixteen scholars

43. Dennis Rohatyn, *Philosophy History Sophistry*

44. Leon Shaskolsky Sheleff, *Social Cohesion and Legal Coercion: A Critique of Weber, Durkheim, and Marx*. Afterword by Virginia Black

45. Alan Soble, Editor, *Sex, Love, and Friendship: Studies of the Society for the Philosophy of Sex and Love, 1977–1992.* A volume **in Histories and Addresses of Philosophical Societies**

46. Peter A. Redpath, *Wisdom's Odyssey: From Philosophy to Transcendental Sophistry*. A volume in **Studies in the History of Western Philosophy**

47. Albert A. Anderson, *Universal Justice: A Dialectical Approach*. A volume in **Universal Justice**

48. Pio Colonnello, *The Philosophy of José Gaos*. Translated from Italian by Peter Cocozzella. Edited by Myra Moss. Introduction by Giovanni Gullace. A volume in **Values in Italian Philosophy**

49. Laura Duhan Kaplan and Laurence F. Bove, Editors, *Philosophical Perspectives on Power and Domination: Theories and Practices*. A volume in **Philosophy of Peace**

50. Gregory F. Mellema, *Collective Responsibility*

51. Josef Seifert, *What Is Life? The Originality, Irreducibility, and Value of Life*. A volume in **Central-European Value Studies**

52. William Gerber, *Anatomy of What We Value Most*

53. Armando Molina, *Our Ways: Values and Character*, Edited by Rem B. Edwards. A volume in **Hartman Institute Axiology Studies**

54. Kathleen J. Wininger, *Nietzsche's Reclamation of Philosophy*. A volume in **Central-European Value Studies**

55. Thomas Magnell, Editor, *Explorations of Value*

56. HPP (Hennie) Lötter, *Injustice, Violence, and Peace: The Case of South Africa*. A volume in **Philosophy of Peace**

57. Lennart Nordenfelt, *Talking About Health: A Philosophical Dialogue*. A volume in **Nordic Value Studies**

58. Jon Mills and Janusz A. Polanowski, *The Ontology of Prejudice*. A volume in **Philosophy and Psychology**

59. Leena Vilkka, *The Intrinsic Value of Nature*

60. Palmer Talbutt, Jr., *Rough Dialectics: Sorokin's Philosophy of Value*, with contributions by Lawrence T. Nichols and Pitirim A. Sorokin

61. C. L. Sheng, *A Utilitarian General Theory of Value*

62. George David Miller, *Negotiating Toward Truth: The Extinction of Teachers and Students*. Epilogue by Mark Roelof Eleveld. A volume in **Philosophy of Education**

63. William Gerber, *Love, Poetry, and Immortality: Luminous Insights of the World's Great Thinkers*

64. Dane R. Gordon, Editor, *Philosophy in Post-Communist Europe*. A volume in **Post-Communist European Thought**

65. Dane R. Gordon and Józef Niznik, Editors, *Criticism and Defense of Rationality in Contemporary Philosophy*. A volume in **Post-Communist European Thought**

66. John R. Shook, *Pragmatism: An Annotated Bibliography, 1898-1940*. With contributions by E. Paul Colella, Lesley Friedman, Frank X. Ryan, and Ignas K. Skrupskelis

67. Lansana Keita, *The Human Project and the Temptations of Science*

68. Michael M. Kazanjian, *Phenomenology and Education: Cosmology, Co-Being, and Core Curriculum*. A volume in **Philosophy of Education**

69. James W. Vice, *The Reopening of the American Mind: On Skepticism and Constitutionalism*

70. Sarah Bishop Merrill, *Defining Personhood: Toward the Ethics of Quality in Clinical Care*

71. Dane R. Gordon, *Philosophy and Vision*

72. Alan Milchman and Alan Rosenberg, Editors, *Postmodernism and the Holocaust*. A volume in **Holocaust and Genocide Studies**

73. Peter A. Redpath, *Masquerade of the Dream Walkers: Prophetic Theology from the Cartesians to Hegel*. A volume in **Studies in the History of Western Philosophy**

74. Malcolm D. Evans, *Whitehead and Philosophy of Education: The Seamless Coat of Learning*. A volume in **Philosophy of Education**

75. Warren E. Steinkraus, *Taking Religious Claims Seriously: A Philosophy of Religion*, edited by Michael H. Mitias. A volume in **Universal Justice**

76. Thomas Magnell, Editor, *Values and Education*

77. Kenneth A. Bryson, *Persons and Immortality*. A volume in **Natural Law Studies**

78. Steven V. Hicks, *International Law and the Possibility of a Just World Order: An Essay on Hegel's Universalism*. A volume in **Universal Justice**

79. E. F. Kaelin, *Texts on Texts and Textuality: A Phenomenology of Literary Art*, Edited by Ellen J. Burns

80. Amihud Gilead, *Saving Possibilities: A Study in Philosophical Psychology*. A volume in **Philosophy and Psychology**

81. André Mineau, *The Making of the Holocaust: Ideology and Ethics in the Systems Perspective*. A volume in **Holocaust and Genocide Studies**

82. Howard P. Kainz, *Politically Incorrect Dialogues: Topics Not Discussed in Polite Circles*

83. Veikko Launis, Juhani Pietarinen, and Juha Räikkä, Editors, *Genes and Morality: New Essays*. A volume in **Nordic Value Studies**

84. Steven Schroeder, *The Metaphysics of Cooperation: A Study of F. D. Maurice*

85. Caroline Joan ("Kay") S. Picart, *Thomas Mann and Friedrich Nietzsche: Eroticism, Death, Music, and Laughter*. A volume in **Central-European Value Studies**

86. G. John M. Abbarno, Editor, *The Ethics of Homelessness: Philosophical Perspectives*

87. James Giles, Editor, *French Existentialism: Consciousness, Ethics, and Relations with Others*. A volume in **Nordic Value Studies**

88. Deane Curtin and Robert Litke, Editors, *Institutional Violence*. A volume in **Philosophy of Peace**

89. Yuval Lurie, *Cultural Beings: Reading the Philosophers of Genesis*

90. Sandra A. Wawrytko, Editor, *The Problem of Evil: An Intercultural Exploration.* A volume in **Philosophy and Psychology**

91. Gary J. Acquaviva, *Values, Violence, and Our Future.* volume in **Hartman Institute Axiology Studies**

92. Michael R. Rhodes, *Coercion: A Nonevaluative Approach*

93. Jacques Kriel, *Matter, Mind, and Medicine: Transforming the Clinical Method*

94. Haim Gordon, *Dwelling Poetically: Educational Challenges in Heidegger's Thinking on Poetry*. A volume in **Philosophy of Education**

95. Ludwig Grünberg, *The Mystery of Values: Studies in Axiology*, edited by Cornelia Grünberg and Laura Grünberg

96. Gerhold K. Becker, Editor, *The Moral Status of Persons: Perspectives on Bioethics*. A volume in **Studies in Applied Ethics**

97. Roxanne Claire Farrar, *Sartrean Dialectics: A Method for Critical Discourse on Aesthetic Experience*

98. Ugo Spirito, *Memoirs of the Twentieth Century*. Translated from Italian and edited by Anthony G. Costantini. A volume in **Values in Italian Philosophy**

99. Steven Schroeder, *Between Freedom and Necessity: An Essay on the Place of Value*

100. Foster N. Walker, *Enjoyment and the Activity of Mind: Dialogues on Whitehead and Education*. A volume in **Philosophy of Education**

101. Avi Sagi, *Kierkegaard, Religion, and Existence: The Voyage of the Self*. Translated from Hebrew by Batya Stein

102. Bennie R. Crockett, Jr., Editor, *Addresses of the Mississippi Philosophical Association*. A volume in **Histories and Addresses of Philosophical Societies**

103. Paul van Dijk, *Anthropology in the Age of Technology: The Philosophical Contribution of Günther Anders*

104. Giambattista Vico, *Universal Right*. Translated from Latin and Edited by Giorgio Pinton and Margaret Diehl. A volume in **Values in Italian Philosophy**

105. Judith Presler and Sally J. Scholz, Editors, *Peacemaking: Lessons from the Past, Visions for the Future*. A volume in **Philosophy of Peace**

106. Dennis Bonnette, *Origin of the Human Species*. A volume in **Studies in the History of Western Philosophy**

107. Phyllis Chiasson, *Peirce's Pragmatism: The Design for Thinking*. A volume in **Studies in Pragmatism and Values**

108. Dan Stone, Editor, *Theoretical Interpretations of the Holocaust*. A volume in **Holocaust and Genocide Studies**

109. Raymond Angelo Belliotti, *What Is the Meaning of Human Life?*

110. Lennart Nordenfelt, *Health, Science, and Ordinary Language*, with Contributions by George Khushf and K. W. M. Fulford

111. Daryl Koehn, *Local Insights, Global Ethics for Business*. A volume in **Studies in Applied Ethics**

112. Matti Häyry and Tuija Takala, Editors, *The Future of Value Inquiry*. A volume in **Nordic Value Studies**

113. Conrad P. Pritscher, *Quantum Learning: Beyond Duality*

114. Thomas M. Dicken and Rem B. Edwards, *Dialogues on Values and Centers of Value: Old Friends, New Thoughts*. A volume in **Hartman Institute Axiology Studies**

115. Rem B. Edwards, *What Caused the Big Bang?* A volume in **Philosophy and Religion**

116. Jon Mills, Editor, *A Pedagogy of Becoming*. A volume in **Philosophy of Education**

117. Robert T. Radford, *Cicero: A Study in the Origins of Republican Philosophy*. A volume in **Studies in the History of Western Philosophy**

118. Arleen L. F. Salles and María Julia Bertomeu, Editors, *Bioethics: Latin American Perspectives*. A volume in **Philosophy in Latin America**

119. Nicola Abbagnano, *The Human Project: The Year 2000*, with an Interview by Guiseppe Grieco. Translated from Italian by Bruno Martini and Nino Langiulli. Edited with an introduction by Nino Langiulli. A volume in **Studies in the History of Western Philosophy**

120. Daniel M. Haybron, Editor, *Earth's Abominations: Philosophical Studies of Evil*. A volume in **Personalist Studies**

121. Anna T. Challenger, *Philosophy and Art in Gurdjieff's Beelzebub: A Modern Sufi Odyssey*

122. George David Miller, *Peace, Value, and Wisdom: The Educational Philosophy of Daisaku Ikeda*. A volume in **Daisaku Ikeda Studies**

123. Haim Gordon and Rivca Gordon, *Sophistry and Twentieth-Century Art*

124. Thomas O Buford and Harold H. Oliver, Editors, *Personalism Revisited: Its Proponents and Critics*. A volume in **Histories and Addresses of Philosophical Societies**

125. Avi Sagi, *Albert Camus and the Philosophy of the Absurd*. Translated from Hebrew by Batya Stein

126. Robert S. Hartman, *The Knowledge of Good: Critique of Axiological Reason*. Expanded translation from the Spanish by Robert S. Hartman. Edited by Arthur R. Ellis and Rem B. Edwards. A volume in **Hartman Institute Axiology Studies**

127. Alison Bailey and Paula J. Smithka, Editors, *Community, Diversity, and Difference: Implications for Peace*. A volume in **Philosophy of Peace**

128. Oscar Vilarroya, *The Dissolution of Mind: A Fable of How Experience Gives Rise to Cognition*. A volume in **Cognitive Science**

129. Paul Bube and Jefferey Geller, Editors, *Conversations with Pragmatism*. A volume in **Studies in Pragmatism and Values**

130. Richard Rumana, *Richard Rorty: An Annotated Bibliography of Secondary Literature*. A volume in **Studies in Pragmatism and Values**

131. Stephen Schneck, *Max Scheler's Acting Persons: New Perspectives*. A volume in **Personalist Studies**

132. Michael Kazanjian, *Learning Values Lifelong: From Inert Ideas to Wholes*. A volume in **Philosophy of Education**

133. Rudolph Alexander Kofi Cain, *Alain Leroy Locke: Race, Culture, and the Education of African American Adults.* A Volume in **African American Philosophy**

134. Werner J. Krieglstein, *Compassion: A New Philosophy of the Other.*

135. Robert N. Fisher, Daniel T. Primozic, Peter A. Day, and Joel A. Thompson, Editors, *Suffering, Death, and Identity*. A volume in **Personalist Studies**

136. Steven Schroeder, *Touching Philosophy, Sounding Religion, Placing Education.* A volume in **Philosophy of Education**

137. Guy Debrock, *Process Pragmatism: Essays on a Quiet Philosophical Revolution*. A volume in **Studies in Pragmatism and Values**

138. Lennart Nordenfelt and Per-Erik Liss, Editors, *Dimensions of Health and Health Promotion.*

139. Amihud Gilead, *Singularity and Other Possibilities: Panenmentalist Novelties.*

140. Samantha Mei-che Pang, *Nursing Ethics in Modern China: Conflicting Values and Competing Role Requirements*. A volume in **Studies in Applied Ethics**

141. Christine M. Koggel, Allannah Furlong, and Charles Levin, Editors, *Confidential Relationships: Psychoanalytic, Ethical, and Legal Contexts*. A volume in **Philosophy and Psychology**

142. Peter A. Redpath, Editor, *A Thomistic Tapestry: Essays in Memory of Étienne Gilson.* A volume in **Gilson Studies**

143. Deane-Peter Baker and Patrick Maxwell, Editors, *Explorations in Contemporary Continental Philosophy of Religion*. A volume in **Philosophy and Religion**

144. Matti Häyry and Tuija Takala, Editors, *Scratching the Surface of Bioethics*. A volume in **Values in Bioethics**

145. Leonidas Donskis, *Forms of Hatred: The Troubled Imagination in Modern Philosophy and Literature*

146. Andreea Deciu Ritivoi, Editor, *Interpretation and Its Objects: Studies in the Philosophy of Michael Krausz*

147. Herman Stark, *A Fierce Little Tragedy: Thought, Passion, and Self-Formation in the Philosophy Classroom*. A volume in **Philosophy of Education**

148. William Gay and Tatiana Alekseeva, Editors, *Democracy and the Quest for Justice: Russian and American Perspectives.* A volume in **Contemporary Russian Philosophy**

149. Xunwu Chen, *Being and Authenticity*

150. Hugh P. McDonald, *Radical Axiology: A First Philosophy of Values*

151. Dane P. Gordon and David C. Durst, Editors, *Civil Society in Southeast Europe*. A Volume in **Post-Communist European Thought**

152. John Ryder and Emil Višňovský, Editors, *Pragmatism and Values: The Central European Pragmatist Forum, Volume One*. A volume in **Studies in Pragmatism and Values**

153. Messay Kebede, *Africa's Quest for a Philosophy of Decolonization*

154. Steven M. Rosen, *Dimensions of Apeiron: A Topological Phenomenology of Space, Time, and Individuation*. A volume in **Philosophy and Psychology**